Economic Aspects of
Animal Breeding

Economic Aspects of
Animal Breeding

Joel Ira Weller

Institute of Animal Sciences
ARO, The Volcani Center
Bet Dagan
Israel

CHAPMAN & HALL

London · Glasgow · New York · Tokyo · Melbourne · Madras

Published by Chapman & Hall, 2–6 Boundary Row, London SE1 8HN, UK

Chapman & Hall, 2–6 Boundary Row, London SE1 8HN, UK

Blackie Academic & Professional, Wester Cleddens Road, Bishopbriggs, Glasgow G64 2NZ, UK

Chapman & Hall Inc., One Penn Plaza, 41st Floor, New York, NY 10119, USA

Chapman & Hall Japan, Thomson Publishing Japan, Hirakawacho Nemoto Building, 6F, 1-7-11 Hirakawa-cho, Chiyoda-ku, Tokyo 102, Japan

Chapman & Hall Australia, Thomas Nelson Australia, 102 Dodds Street, South Melbourne, Victoria 3205, Australia

Chapman & Hall India, R. Seshadri, 32 Second Main Road, CIT East, Madras 600 035, India

First edition 1994

© 1994 Joel Ira Weller

Printed in Great Britain by TJ Press (Padstow) Ltd, Padstow, Cornwall

ISBN 0 412 59750 0

A catalogue record for this book is available from the British Library

To my teachers Rom Moav and Morris Soller
and to my wife Hedva

Contents

PREFACE

"The seven good cows are seven years;
and the seven good ears are seven years:
the dream is one. "

Genesis XLI:26

When I started working on the subject of economic aspects of animal breeding, I was reminded of the story of a woman walking down the street complaining, "My son-in-law doesn't know how to play cards! He doesn't know how to play cards!"

Finally someone stops her and asks, "What's so terrible about that?"

The woman answers, "The problem is that he does!"

To a large extent we animal breeders have been like the son-in-law in the story. We have been busy breeding even though we do not know how to do it. Many sophisticated readers may object to this analogy. They can argue that we have a good understanding of the principles of genetics, and even if we do not know how to apply these principles directly to many traits of interest, we can still use statistics and quantitative genetics to change population values for these traits. It can further be argued that not only do we know how to breed, we have been doing it successfully for quite a while. During the last fifty years, great progress has been made in various domestic species; cows give more milk, poultry grow faster, and give more eggs, etc. Nevertheless, the question still remains: "True we are doing something, but are we doing the right thing?"

Generally breeders have tried to breed for clearly defined, obtainable goals without regard as to whether these goals are desirable. In the example of dairy cattle, there is a wealth of literature dealing with how to breed for increased milk production, but very few papers address the question of whether or not we should be doing this.

Many geneticists and breeders will explain this discrepancy by saying that since breeding goals are generally determined by economic considerations, this problem should be left to economists. Convinced of the veracity of this statement, I went to my university library, and took out half a dozen books on agricultural economics. None of them mentioned breeding or genetics. Pursuing this question further in the literature, I became convinced that, whether or not economists should by right be pursuing this question, the fact is that they are not. The vast majority of work done on determining breeding goals has been done by geneticists, who have acquired some training in economics (present company included).

Another objection that the sophisticated readers may raise is that breeding goals are often rigidly determined by factors beyond our control. Clearly dogs and cats are bred for certain characteristics that have no economic importance, other than the fact that people are willing to pay for them. Even with farm animals, farmers in many countries will prefer animals they consider attractive, regardless of the effect of these traits on economic performance. These arguments still do not contradict the need for determination of breeding goals within an economic context. Although a cow's color may have no effect on her ability to produce milk, if someone is willing to pay more for red cows, this trait, by definition, has economic importance.

A final objection that my hypothetical reader could raise to my opening analogy is that the problem has been solved long ago. In 1943 L. N. Hazel published "The genetic basis for constructing selection indexes" in *Genetics*. This classic paper, which is the basis for economic selection index provided the methodology for determining selection goals based on the relative economic values of the traits under consideration. Again here we see an anomaly when we survey the literature. Although the principles of selection index have been known for over forty years, relatively few scientific studies, or commercial enterprises have actually attempted to apply them in the method envisioned in Hazel's paper. One reason for this is that the basic principles of selection index are not a complete solution to the problem of determination of selection goals. (I should mention that I have no intention to belittle the importance of this paper, which will be the basis for a large part of this book.) Without going into detail at this point, I would just mention that economic selection index assumes that the relative economic weights of the traits under consideration are known. If this is not the case, then traditional selection index is of no use in determination of breeding objectives.

The goals of this book are to review and organize the literature dealing with economic evaluation of genetic differences, determination of breeding goals within an economic context, and economic evaluation of breeding programs. Although nearly all of the material presented can be found in scientific literature, it is scattered through a number of different journals and symposia. To the best of my knowledge no similar work on the subject exists. There are a few review articles that deal with certain aspects of the general topic, and these will be discussed in detail, but no review article deals with the entire question.

One difficulty I encountered in summarizing the literature was the inconsistency of symbols used by different authors, or even the same authors in different papers. I have therefore made a supreme effort throughout to be consistent in the use of symbols, and not to use the same symbol to represent two different quantities. I used the convention of denoting vectors in lower case **bold** type, and matrices in upper case **bold** type. Scalars were denoted by both lower and uppercase regular type symbols. Where possible I tried to accept the most commonly used terminology. It was nevertheless necessary to alter the symbols used in many papers to conform to the standard I adopted. Unfortunately, in a few cases, I even had to alter the symbols for quantities presented in a single literature source because of a conflict with symbols used elsewhere in this book.

To aid the reader, an alphabetical glossary of symbols is presented after the text.

Even though this book is entitled *Economic Aspects of Animal Breeding*, most of the material discussed is also relevant to plant breeding. The reason for the limitation in the title is that, even though most of the problems discussed are applicable to all domestic species, these problems are generally more acute in animal breeding. Therefore although there are a few mentions of problems specific to plant breeding, nearly all the examples brought will deal with domestic animals. *Economic Aspects of Animal Breeding* is written on the level of a graduate student in genetics or animal breeding. Established scientists may find some of the material unnecessary.

The book is divided into five parts. Part I deals with basic concepts necessary for dealing with the questions of interest. This part includes very little material not found in other texts on the specific topics covered. A more advanced reader may wish to skip some or all of this part. Part II, which deals with economic evaluation of genetic differences, includes material which, although present in the literature, might be unfamiliar to many geneticists and commercial breeders. Part III on advanced topics in selection index, requires a reasonable familiarity with matrix algebra. The more casual readers may wish to skip chapter 9, which is highly mathematical. Part IV, on the economic evaluation of breeding programs contains a significant quantity of original material, because the literature is relatively weak on this topic. This part also includes economic aspects of new advances in biotechnology. The final part deals with crossbreeding and heterosis, which is of importance in most domestic species.

I would first thank all the authors whose work I have included in *Economic Aspects of Animal Breeding*. Chief among them is my teacher, the late Dr. Rom Moav, who first interested me in this topic, and is still the major source for this text. If his life had not been tragically cut short, he might have written this book. As Isaac Newton said, "If I can see far, it is only because I stand on the shoulders of giants." I also thank my teacher, Dr. Morris Soller, Rom Moav's colleague, who encouraged me to begin this undertaking, and my wife Hedva who gave me the support necessary to finish. I also wish to thank Michael Grossman for his encouragement, and Suzanne M. Hubbard for helping through the intricacies of modern word processing and graphics. Finally I thank Ephraim Ezra for encouragement, and useful suggestions throughout.

Joel Ira Weller
Rehovot, Israel
August, 1993
Elul, 5753

PART I

BASIC CONCEPTS

This section summarizes basic concepts in quantitative genetics, economics, matrix algebra, and systems analysis, necessary for understanding the main topics of this book. As stated in the Preface, very little material not found in other texts is presented, and readers familiar with the topics covered may choose to skip the relevant chapters. On the other hand, readers not familiar with the concepts presented may wish additional reading. The texts upon which I relied to write this section, and which I recommend for additional information were: *Quantitative Genetics* (Falconer, 1964); *Economics* (Samuelson, 1980; and *Matrix Algebra Useful in Statistics* (Searle, 1982). I was unable to locate a suitable text on systems analysis, but for the interested reader I would recommend *The Study of Agricultural Systems* (Dalton, 1975) which although not a text, is the most relevant work on the topic of which I am aware.

Chapter One

Basic Concepts in Quantitative Genetics

1.1 Introduction

The principles of quantitative genetics have served until the present as the basis for most genetic improvement of domestic species by human manipulation. These principles will be summarized in this chapter. Rather than deal with individual genes, quantitative genetics deals chiefly with trait values, and statistics of these values derived from populations of interest. The most important statistics are the arithmetic mean and the variance. In the case of the normal distribution, these two statistics define the population in question. We will assume that the reader has an understanding of the basic principles of Mendelian genetics and statistics. Therefore, relatively simple statistic terms, such as normal distribution, mean, variance, standard deviation, correlation, and regression will not be defined.

1.2 Quantitative vs. categorical traits

Shortly after Gregor Mendel's results in pea plants were rediscovered in 1900, (Mendel, 1866; Peters, 1959) the question arose as to whether these results were applicable to most traits of economic importance in domestic species. Traits such as milk yield, fruit weight, or growth weight did not seem to have the simple distributions of values that Mendel found in the traits he studied. Instead, nearly all of these traits had close to normal distributions, or could be converted to normal distributions by relatively simple transformations. Furthermore, experiments conducted in the early part of this century showed that although there generally was some similarity between parents and offspring, it did not seem to conform to the simple Mendelian rules (Johannsen, 1903). These traits were termed quantitative traits, and their genetic analysis was termed quantitative genetics.

It soon became clear that if the phenotypic value for these traits was determined by a number of Mendelian loci segregating independently, then the statistical distribution for the trait within the population would approach normality (Fisher, 1930). This would be true whether the mode of inheritance of the individual loci was dominant or codominant. The greater the environmental effect on the trait in question, the lower the number of loci necessary to obtain a nearly normal distribution. However, even if the number of loci affecting a quantitative trait was low, determination of the effect of the individual loci (termed polygenes) would be difficult, if not close to impossible. It was argued that the important questions on quantitative traits from a breeding point of view, such as optimization of selection programs, genetic evaluation of animals, and prediction of the response to selection, could be answered without knowledge of the molecular mode of inheritance. Thus even though the Mendelian rules were correct on the underlying level, they were generally irrelevant.

1.3 Phenotype vs. genotype, environmental and genetic variation

Mendel, through his experiments, understood that there was a difference between phenotype (the observed expression of a given trait) and genotype (the individual's genetic component for the trait). He demonstrated that plants with the same phenotypic value could have different genotypes. Since the genotypes for quantitative traits are determined by a number of loci, a major goal of quantitative genetics is to estimate what fraction of the trait in question is determined by the individual's genotype. Algebraically this relationship can be written as follows:

$$x = g + e \qquad\qquad [1.1]$$

where x is the observed or phenotypic trait value for the individual, g is the individual's genetic value, and e is the difference between x and g, which includes all non-genetic factors that determine an individual's observed trait value. To distinguish from the genetic effect included in g, e will be termed the environmental effect. Since there is clearly a mathematical dependency between g and e, we will assume throughout that all factors are measured as deviations from the population mean. For example a particular cow may give 9000 kg milk/yr while the population mean is 8000. Then x will equal 1000 kg/yr. Assuming we know that $g = 300$ kg/yr, we can then deduce that the environmental effect for this cow is 700 kg/yr.

Although equation [1.1] seems trivial, it is in fact the basis for determination of the genetic values of individual animals, which is the basis for genetic selection. Since breeding is our major concern, the main question of interest is determination of g. Of course once this is done we will also know e. We will start by assuming that both g and e are normally distributed. This is generally a reasonable assumption for e, since that part of the observed value not determined by genetics will probably be determined by a large number of more-or-less independent factors. (It can readily be shown that the distribution of any trait derived as the sum of a large number of relatively small factors, will tend to normality, regardless of the distributions of the individual factors.) For g the assumption of normality will also hold, unless the genotype is chiefly determined by a very few major genes.

How then do we distinguish between g and e? One method that comes to mind is to try to control e. For example, give all individuals the same environment. Thus all differences between individuals will be due to g. Although in theory correct, this is nearly impossible in practice. Not all environmental factors that can affect quantitative traits are known, and even some known factors are beyond our control. Certain factors which at first glance may seem of no importance are sometimes found to have major effects on traits of economic importance, such as position of a chick in the hen house. For species with vegetative reproduction this problem can be solved by raising large numbers of genetically identical individuals. Over a large sample, it can be assumed that the mean environmental effect is zero, and any deviation of the isogenic sample mean from the population mean can be ascribed to g. This is in fact commonly done for many commercial plants, but is of course not possible for most animal species of economical importance.

1.4 Additive and non-additive genetic variation

For animals, the main method for estimation of g will be by comparison between relatives. For instance, a sire or a dam passes on half of its genes to its offspring. However, this type of analysis is also more complicated than it appears at first glance. One pitfall with this type of analysis can be explained by going back to the results of Mendel. The loci that Mendel studied displayed complete dominance. Thus if both parents were homozygous, one for the recessive and the other for the dominant allele, the offspring would be similar only to the parent with the dominant allele. Even in this situation, there will still be a positive correlation between the phenotype of parents and their progeny over the whole population. To deal with this and similar problems, quantitative genetics have defined two types of genetic effects, additive and non-additive. Algebraically the relationship can be written as follows:

$$g = g_a + g_b \qquad\qquad\qquad [1.2]$$

Where g_a is the additive portion of an individual's genotype, g_b is the non-additive portion, and g the individual's genetic value, as defined previously. Included in the additive portion of an individual genotype will be those factors that determine the similarity between parents and progeny, while g_b includes factors in the individual's genotype which will not affect the phenotype of the progeny in an additive manner. We have already seen that for loci with a dominant-recessive mode of inheritance, part of the genetic component will behave in a non-additive manner. On a statistical level this can be considered an interaction between the two alleles of the loci.

Although Mendel in his experiments found simple dominant-recessive relationships for the loci he studied, this is not necessarily the case for all loci. A situation of partial dominance is also possible. For example, assume that the mean value of a trait in a population is 100 units, and that a given locus with two alleles, A and a, affects this trait. Assume further that individuals with the genotype AA have a mean value of 110 units, and individuals with the genotype aa have a mean value of 90. If this locus behaves in the classical Mendelian fashion then individuals with genotype Aa will also have a mean value of 110. We will now define three alternative possibilities:

1. Codominance: individuals with genotype Aa have a mean value equal to the mean of the two homozygotes. In this situation the gene will contribute only to the additive component of genetic variance.
2. Partial dominance: individuals with genotype Aa will have a mean between the homozygote midpoint and one of the homozygotes. In this case, as in the case of complete dominance, this gene will contribute to both the additive and non-additive genetic components.
3. Overdominance: the mean value of the heterozygote is outside the range of the two homozygotes. In the extreme case, where the two homozygotes have equal value, then this locus will contribute only to the non-additive component of genetic variance.

Just as there can be intralocus interactions, there can also be interloci interactions. Genetics term interactions between loci "epistasis". Dominance and epistasis are the two factors that account for non-additive genetic effects. Thus equation [1.2] can be expanded as follows:

$$g = g_a + g_d + g_z \qquad\qquad [1.3]$$

where g_d and g_z are the individual's deviation from the population mean due to dominance and epistatic effects, respectively. In most domestic animals, it is very difficult to estimate $g_d + g_z$.

Since the additive genetic effects explain the similarity between parents and offspring, it is also called the animal's breeding value. Thus for example if a particular dairy cattle sire has a large number of daughters, and these daughters average 500 kg milk above the population, it can be deduced that the sire's breeding value is $2 \times 500 = 1000$. The mean daughter difference is multiplied by two because only half of the sire's genes are transmitted to each daughter.

For this reason, the sire evaluations published in some countries, including the US, are termed "Predicted Differences", or "Predicted transmitting abilities", which are equal to one half of the animal's breeding value.

1.5 Variance components, heritability, and repeatability

For any trait of interest, a central question will be how important are the various components that determine the individual's phenotype. In other words, how much of the total variation in the population is explained by g_A, g_D, g_Z, and e? From basic statistics we know that for any two variables, x and y, if:

$$z = x + y \qquad\qquad\qquad [1.4]$$

then:

$$\sigma_z^2 = \sigma_x^2 + \sigma_y^2 + 2\sigma_{xy} \qquad\qquad\qquad [1.5]$$

where σ_z^2, σ_x^2, and σ_y^2 are the variance of z, x, and y, respectively, and σ_{xy} is the covariance between x and y. Thus from equations [1.1] and [1.5] we can deduce the following equality:

$$\sigma_x^2 = \sigma_g^2 + \sigma_e^2 + 2\sigma_{ge} \qquad\qquad\qquad [1.6]$$

where σ_x^2, σ_g^2, and σ_e^2 are the variances for x, g, and e, and σ_{ge} is the covariance between g and e. We will generally assume that σ_{ge} is zero, even though in many practical situations this is not the case. For example, farmers may vary the amount of feed given to individual cows on the basis of the cow's sire and dam. This will generate a positive covariance between the environment (feeding level) and the genome (sire and dam genetic level). σ_g^2 and σ_e^2 will therefore be termed the variance components of σ_x^2. Based on equation [1.3], and again assuming that all covariances are equal to zero, the phenotypic variance can be decomposed as follows:

$$\sigma_x^2 = \sigma_A^2 + \sigma_D^2 + \sigma_Z^2 + \sigma_e^2 \qquad\qquad\qquad [1.7]$$

where σ_A^2, σ_D^2, and σ_Z^2 are the additive, dominance, and epistatic genetic components of variance, and the other terms are as defined previously. Since, as stated previously, the additive effects determine similarities between relatives, we will be most interested with σ_A^2. Although it is possible to estimate the other components of variance in equation [1.7] for specific population structures, it will almost always be more difficult than estimation of σ_A^2 and is generally not done for animal populations.

now do geneticists estimate σ_A^2? Two methods are generally used; parent-offspring regressions, and analysis of variance. The first case can be illustrated by assuming that we have milk records on a number of dam-daughter pairs. We then compute the regression of the daughter on her dam. The following equation can be described:

$$b_{op} = \sigma_{op}/\sigma_x^2 \qquad [1.8]$$

where b_{op} is the regression of offspring on her dam, σ_{op} is the parent-offspring covariance, and σ_x^2 is the phenotypic variance, as defined above. Assuming that there are no sources of similarity between daughters and dams except for additive genetic variance, then σ_{op} will be equal to one half of the σ_A^2, since, as stated above, a parent passes one half of its genome to its progeny. Thus σ_A^2 can be estimated as follows:

$$\sigma_A^2 = 2b_{op}\ \sigma_x^2 \qquad [1.9]$$

For example assume that σ_x^2 for annual milk production = 1,000,000 kg, and b_{op} = 0.12. Then σ_A^2 = 240,000 kg.

To estimate σ_A^2 by analysis of variance, assume that we have a population consisting of a number of sires, each with a relatively large number of daughters. Assume further that each sire was mated to a random sample of dams, and that all environmental effects are randomly distributed. If these conditions are true, we can then assume that the between-sire component of variance from an ANOVA analysis will consist only of additive genetic effects. Since as stated previously, a parent contributes only half of its genome to its progeny, the between-sire component of variance will be $1/4\ \sigma_A^2$. (Remember that variances are differences squared. Therefore the 1/2 effect of the genome that is passed must be squared on the level of variance components.) Thus if the between-sire component of variance for annual milk production is 60,000 kg, σ_A^2 = 4 × 60,000 = 240,000 kg, the same result as presented above.

Since different traits are measured in different units, σ_A^2 values cannot be compared across traits. For this and other purposes, heritability (h^2), a unitless statistic is defined as follows:

$$h^2 = \sigma_A^2/\sigma_x^2 \qquad [1.10]$$

It might strange that the symbol for heritability is "h^2" and not h, especially since the square root of heritability is not as important a statistic. This symbol is derived from Wright's 1921 terminology, and is universally accepted. Clearly the same methods described above to estimate g_A can also be used to estimate heritability. We will now rewrite equation [1.8] in terms of heritability:

$$b_{op} = \sigma_{op}/\sigma_x^2 = \sigma_A^2/(2\sigma_x^2) = h^2/2 \qquad\qquad [1.11]$$

Thus in the example above where $b_{op} = 0.12$, $h^2 = 0.24$, h^2 can also be derived as $\sigma_A^2/\sigma_x^2 = 240,000/1,000,000 = 0.24$, the same value.

Sometimes it is possible to sample the phenotype several times on the same individual, each time with a different environmental component. Examples are milk production of a cow over several lactations, or the number of piglets in a litter over several litters. In these cases it is of interest to estimate how much of the variance is due to genetics and the permanent environmental effects, vs. environmental effects which will differ for each expression of the trait. In essence we are asking how much is this individual "worth" for a given trait. This statistic is called "repeatability" and is defined as follows:

$$rpt = (\sigma_g^2 + \sigma_{PE}^2)/\sigma_x^2 \qquad\qquad [1.12]$$

where rpt is repeatability, σ_{PE}^2 is the permanent environmental effect, and the other terms are as defined previously. Note that the numerator includes σ_g^2, all the genetic variance, and not just σ_A^2. Since we are interested in this case in the phenotypic expression of the trait in common, all the genetic variance, including the non-additive portion, is included in the numerator. As an example we can again take dairy cattle, where repeatability for milk production has been estimated at 0.5. Remember that heritability for this trait is only about 0.25. Of course repeatability will always be greater than or equal to heritability, and for many traits the difference is quite significant.

Unfortunately, this same term, "repeatability", has been defined differently in some of the quantitative genetics literature. In the hope of reducing the confusion, which already exists, the second definition of repeatability will be presented shortly.

1.6 Estimation of breeding values

Until now we have been describing the genetic and phenotypic parameters of populations. The main goal of agricultural genetics is to modify the genetic parameters of populations of interest. (Throughout we will consider mainly the goal of changing population means, although changing variances is also sometimes important and will be considered later.) The main vehicle for changing the mean genetic value will be selection, i.e. progeny of individuals of the desired genotype will be kept for reproductive purposes, while individuals with undesirable genotypes will not reproduce, or their progeny will be discarded. The two central questions in genetic selection are: the determination of candidates for selection, and estimation of the expected response to selection.

As stated above, there is no method at present to directly determine the genetic value of an individual. If we look at equation [1.1] we see that two animals with equal genetic values can have different phenotypic values due to environmental factors. Thus the "best" individual phenotypically is not necessarily the "best" genetically. However if the covariance between genetic and environmental effects is insignificant, then there will be a positive correlation between genetic and phenotypic values. From equation [1.8] and [1.9] we see that if g_A is greater than zero, there will be a positive regression of offspring performance on parental performance for the trait of interest, equal to half of the heritability. Using this regression constant we can then write the following equation:

$$Y_{op} = b_{op}(Y_P) \qquad [1.13]$$

where Y_{op} is the predicted trait value for a progeny of a given parent, Y_P is the trait value of the parent. As stated previously, we will assume that both Y_{op} and Y_P are measured as deviations from the population mean. As defined above in section 1.3, Y_{op} is the predicted difference of the parent, which is equal to 1/2 of its breeding value. Thus the following equation can also be formulated.

$$BV_P = 2b_{op}(Y_P) = h^2(Y_P) \qquad [1.14]$$

where BV_P is the parent's breeding value. Thus heritability is also the regression of breeding value on phenotypic value.

In many practical breeding situations, an individual is evaluated on the basis of the phenotypic values of a number of related individuals. In this case, breeding values can be estimated by a methodology called "selection index". (In the literature, the term, "selection index," is used both for the methodology described above, and for determination of economic weighting of individual traits in an overall breeding objective. To distinguish between the two, we will denote the former "genetic selection index", and the latter "economic selection index".) In the simple case where the breeding value of a single individual is estimated from a single quantity, selection index can be formulated as follows:

$$BV_i = b(x_j) \qquad [1.15]$$

where BV_i is the estimated breeding value for individual i, b is a regression constant, x_j is the phenotypic deviation from the population mean on the group of related individuals j. As for any regression, b can be calculated as the covariance of BV_i and x_j, divided by the variance of x_j. For example, a common situation is the evaluation of a sire on the basis of the records of a number of progeny. In this case the sire's breeding value can be estimated as a regression on the mean of progeny records, $Y_P.$, as follows:

$$BV_P = \frac{(1/2)\sigma_A^2}{(1/4)\sigma_A^2 + [(3/4)\sigma_A^2 + \sigma_e^2]/n} (Y_P.) \qquad [1.16]$$

where n is the number of progeny, and σ_e^2 now includes all phenotypic variance not included in σ_A^2. This equation is derived as follows: The covariance between the parent's breeding value and the mean of the progeny's records is $(1/2)\sigma_A^2$, since each progeny receives half of its genes from the parent. The variance of $Y_P.$ will be the denominator of the right-hand side of equation [1.16], and is computed as follows: The variance of the mean of a sample will be the variance of the sum of the sample divided by n^2. The variance of the sum will be the sum of the variances and twice all the possible covariances, as shown in equation [1.5] for a sample of two observations. The covariance among progeny will be $1/4\sigma_A^2$, since each offspring receives a random one half of its genes from the common parent. It can be shown that after summing and combining these variances and covariances, the denominator of equation [1.16] is obtained. In terms of heritability equation [1.16] can be rewritten as follows:

$$BV_P = \frac{(1/2)h^2}{(1/4)h^2 + (1 - (1/4)h^2)/n} (Y_P.) \qquad [1.17]$$

In the case of a single progeny, this equation reduces to the same value as equation [1.11], and as n approaches infinity, BVP will approach Y_P. As defined previously the predicted difference of the parent will be equal to one half of the breeding value, that is:

$$PD = \frac{(1/4)h^2}{(1/4)h^2 + (1 - (1/4)h^2)/n} (Y_P.) \qquad [1.18]$$

The regression of PD on $Y_P.$ is also denoted as "repeatability", especially by dairy cattle breeders. This definition should not be confused with the definition of repeatability brought above in Section 1.5, which is the more generally accepted definition. This regression is also equal to the coefficient of determination between the evaluation and the individual's true genetic value. If the phenotypic performance of the individual is the only information available, then the coefficient of determination will be equal to the heritability of the trait. In both cases the square root of this quantity will be equal to the correlation between the evaluation and the true genetic value. This correlation is sometimes called the "accuracy" of the evaluation. To minimize confusion in this book henceforth, only the definition of Section 1.5 will be used for repeatability.

Similar to the case just described, selection index formulas can be derived to estimate breeding values for any combination of individuals, based on which relatives have records. However these calculations become quite complicated, unless use is made of matrix algebra. This will be discussed in Chapter 3, where the general framework for both genetic and economic selection index will be formulated.

1.7 Estimation of response to selection

We will now discuss the estimation of the expected response to selection, which we will see is intimately related to the estimation of breeding values. We will start with the simplest situation called "mass selection." In mass selection, we assume that we have a population of individuals which we wish to select for a given trait. Individuals with the highest value for this trait are selected as parents for the next generation. These individuals are then mated among themselves to produce progeny for the next generation. An example of selection of beef cattle for weight at weaning is illustrated in Figure 1.1. In this example, the objective is to increase the mean value for this trait by mass selection. Heritability is assumed to be 0.25. The phenotypic value of each dam and her progeny in a simulated population are plotted. Although the correlation between parents and progeny is evident, there are individual parents with high values who had offspring with low values.

We will select by only keeping progeny of parents with the highest 10% of trait values. This type of selection is called truncation selection, and the truncation value is marked in Figure 1.1. Assuming that the selected individuals have a mean value of S above the general population mean, the mean of their progeny can be estimated as follows:

$$\phi = b_{op}S \qquad [1.19]$$

where ϕ is the difference between the progeny mean and the general mean in the parental generation, and b_{op} is the regression of offspring on the midparent value. ϕ can also be defined as the response to selection. In the case of mass selection, the regression coefficient, b_{op}, will be equal to the heritability. Thus to predict the response to selection, it is necessary in this case to know the heritability and S.

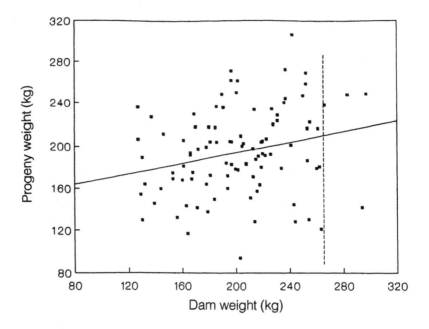

Figure 1.1. Regression of progeny on dam weaning weight for beef cattle. Each point represents the weaning weight of a progeny and her dam. Solid line is the regression. Broken line is the value for truncation selection based on dam weaning weight.

We have already described methods for estimating h^2. However it is not immediately evident how to estimate S. If we select the 10% best cows, how much better will these cows be than the population mean? Clearly, S will depend on the trait units. Therefore we will define a new unitless quantity, i, as follows:

$$i = S/\sigma_P \qquad\qquad [1.20]$$

where i is the "selection intensity", and σ_P is the phenotypic standard deviation. If the population is large and has a near normal distribution, i can be computed as follows:

$$i = z/p \qquad\qquad [1.21]$$

where z is the ordinate of the normal curve, and p is the proportion of individuals selected. Combining equations [1.19] and [1.20], and substituting h^2 for b_{op} for the case of mass selection, we have:

$$\phi = ih^2\sigma_P \qquad [1.22]$$

For example, assume that we are selecting cattle on weaning weight. The heritability of this trait is 0.25, and the phenotypic standard deviation is 40kg. If the highest 10% are selected as parents for the next generation, p = 0.1, and z = 0.18. Therefore, i = 0.18/0.1 = 1.8, and the selection response will be: ϕ = 0.25(0.18)40/0.1 = 18kg. This is the response per generation. Often it will be of interest to compute the response per unit time, which we will denote ΔG. In this case we will define the following equation:

$$\Delta G = \phi/L \qquad [1.23]$$

where L is the generation interval.

Since h^2 is the ratio of genetic to phenotypic variance, equation [1.18] can be rewritten as follows:

$$\phi = ih(\sigma_A) \qquad [1.24]$$

where σ_A is the genetic standard deviation. (Remember that we previously defined "h" as the "accuracy" of the genetic evaluation.) Equation [1.24] is useful if selection is based on genetic evaluations, rather than phenotypic records. As shown previously, for the case of sires evaluated based on their progeny, the accuracy of the evaluation will be the square root of the regression coefficient in equation [1.18].

Finally, we will consider the case of different selection intensities along the different paths of inheritance. For vertebrates there are four paths of inheritance: sires of one generation to sires of the next, dams to dams, dams to sires, and sires to dams. In nearly all selection schemes, both the selection intensities and generation intervals for the four paths will be different. For example, since a cow produces only one calf a year, selection intensity along the dam-to-dam path will be low. Nearly all female calves must be raised for replacement. However, since male fertility is extremely high, especially with artificial insemination, only a few bull calves must be kept for breeding. Therefore the dam-to-sire breeding path can have a very high selection intensity. Under these conditions, ΔG can be computed as follows:

$$\Delta G = \Sigma\phi/\Sigma L \qquad [1.25]$$

that is, the genetic response per unit time will be equal to the sum of responses over all selection paths, divided by the sum of their generation intervals.

1.8 Inbreeding and selection

Inbreeding results from the mating of related individuals, and generally has a negative effect on the fitness of species which are normally outbreeders. The negative effect of inbreeding in normally outbred populations is due to a major increase in frequency of the expression of rare deleterious recessive alleles. For example, if the frequency of a deleterious, recessive allele is 0.01, then the frequency of homozygotes assuming random mating will be $(0.01)^2 = 0.0001$. However, for inbred individuals the frequency may be much higher, because the inbred progeny can receive the same allele from both parents if they are descendant from the same carrier. For example, a progeny from the mating of two half-sibs will have a probability of 1/8 of receiving the same grandparent allele from both parents. This probability is called the "coefficient of inbreeding".

Selection, by limiting the number of individuals used as parents for the next generation, almost always results in an increase in the level of inbreeding in the population. Furthermore, the increase of inbreeding per generation is cumulative. Van Vleck (1981) gives the following formula for the increase in inbreeding, ΔF, per generation:

$$\Delta F = (N_S + N_D)/(8N_SN_D) \tag{1.26}$$

Where N_S and N_D are the number of sires and dams per generation.

Numerous studies have reported negative effects of inbreeding on traits of economic importance, such as growth rate and milk production. In addition, inbreeding, like selection, decreases genetic variance, and so decreases the rate of genetic gain. However, this second problem is generally insignificant relative to the rate of increase in the expression of negative alleles.

1.9 Summary

In this chapter we demonstrated how the basic rules of Mendelian genetics were used to explain the continuous variation found in most traits of economic importance. We showed that even though the expressions of these traits are affected by many genes and environmental effects, predictions useful in breeding can be made, if adequate information is obtained. Only part of the genetic factors passed from one generation to the next will actually explain the similarity between most relatives. Information on the individual itself, and its relatives, can be used to estimate the individual's breeding value, and to predict response to selection.

Several important statistics were introduced in this chapter. The important unitless statistics are heritability (h^2), repeatability (rpt), accuracy (h), and selection intensity (i). The important statistics which are measured in trait units are breeding value (BV), predicted difference (PD), and the response to selection ΔG. Important variance components, which are in trait units squared are the phenotypic (σ_x^2), the additive genetic (σ_A^2), and the environmental (σ_e^2) components of variance.

Chapter Two

Basic Concepts in Economics

2.1 Introduction

Contrary to the previous chapter, which was a general review of quantitative genetics, in this chapter we will cover only those principles of economics relevant to animal breeding. There are many topics in economics that are generally covered in an introductory economics course that have little relevance to the topic at hand, and will therefore not be discussed. The topics we will consider are the production-possibility frontier; the law of diminishing returns; demand and supply curves; equilibrium and competition; elasticity of supply and demand; momentary, short- and long-term equilibrium; application of supply and demand; the dynamic cobweb; marginal cost and utility; fixed and variable costs; interest, discount, and inflation rates, and profit horizon. We will also discuss special topics in agricultural economics, including long-run decline, short-run instability, and government aid. Hopefully at the end of this chapter we will understand why "the price of pig, is something big," (H. J. Davenport) and what breeding can do about it.

2.2 Production-possibility frontier and the law of diminishing returns

A reasonable starting point for the discussion of economics is with the "Law of Scarcity". That is, no matter how affluent a society, no society can produce as many goods and services as people want. This is because resources (labor, capital, raw material, land, etc.) are limited. Those goods of which everyone has as much as he wants, no matter how important they may be, are therefore not economics goods. Thus on this world, air is not an economic good, and has no price. This of course is not the case in a space capsule, where air would be in limited supply, and therefore very valuable.

Since resources necessary to produce economic goods are limited, a society must choose how much of what to produce with the limited means at hand. This problem is often in the news as the choice governments face between "guns or butter." In other words, if more resources of the society are directed to military expenditures (guns), there will be less resources left for other expenditures (butter). This of course will be true for all economic goods and services that society may wish to produce.

This relationship is displayed graphically in Figure 2.1 for a hypothetical society. In this simple example we will assume that this society produces only two goods - guns and butter. We see that the maximum number of guns that this society can produce is 15,000, but this will only be accomplished if all resources are directed to production of guns. Conversely if all resources are directed to butter production, 5,000 tons of butter will be produced, but there will be no guns. If part of the resources are directed to each objective, both guns and butter will be produced, but less of each good than is absolutely possible. The curve that describes the maximum quantity of each good that can be produced, as a function of how resources are allocated, is called the "production-possibility frontier." Although it is possible to produce less than this frontier, if production is inefficient, it is not possible to produce more.

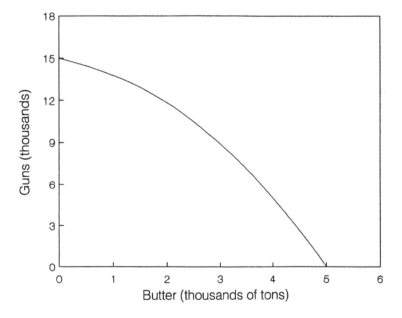

Figure 2.1. The production-possibility frontier for a hypothetical society that produces only two goods - guns and butter.

In Figure 2.1, this curve is concave. This will be the general rule for production-possibility frontiers, and can be explained as follows: if we start with the situation where only butter is produced, and begin to direct resources to the production of guns, maximum total production will be obtained if those resources which are most efficient in the production of guns are directed first in this direction.   It is reasonable that these resources will be less efficient in the production of butter, and thus the total quantity of goods produced will increase. As the production of guns increases, it will be necessary to direct in this direction more resources, which will be less efficient in the production of this commodity. Thus although production of guns will increase, the rate of increase will decline.

Just as a society must choose between production of different goods, it must also choose between current consumption and investment. Through investment, or capital formation, it is possible to change the production-possibility frontier. By investing in new machinery, research and development, etc. a society can move to a situation where it is possible to produce both more guns and butter. Of course, this will be at the expense of less current production and consumption. This consideration is of special concern to animal improvement, since breeding is, by its nature, a long-term procedure.   Animal breeding requires that resources be directed from current consumption to investment in breeding programs. For example, in the breeding of dairy cattle, resources are spent on record keeping and data analysis, so that superior sires can be selected. Future daughters of these sires will be able to produce more butter from the same resources, or the same quantity of butter from less resources. In economic terms the return can be measured in a change in the production-possibility frontier to the right. This situation is illustrated in Figure 2.2.

Similar to the situation in Figure 2.1, the curves in Figure 2.2 are also concave. This means that, starting from a situation where all resources are directed to consumption, direction of a small fraction of all resources to investment will result in a relatively large "return" in increased future production.   However, as more resources are directed to investment, the proportionate return in increased future production will be less. Again this is because those resources which can most efficiently increase future production will be directed first to investment.   The discussion of investment will be continued in section 2.9.

This principle, which explains the relationships in Figures 2.1 and 2.2, is call the "Law of Diminishing Returns". This law can be phrased as follows: as inputs are increased, production will increase, but each additional equal increase in inputs will result in a diminished increase in outputs, as compared to the previous increase.   This law describes the relationship between inputs and outputs, not between alternative inputs or outputs. We have already seen one reason why this is true, namely, that generally speaking, the most efficient resources will be the first directed to production of a given good or service. Another reason is the concept of the limiting factor. This can be explained with the following example. Assume that ten workers take care of 1000 milk cows. For a given management system this number may not be nearly enough, and it is possible that by doubling the number of workers, production would more than double. What would happen if the number of workers are again doubled, this time to forty? With forty workers, it is still possible to feed the cows better and milk them more frequently. However at this point, doubling the number of workers will not double milk production. This is because factors other than labor are now limiting production, specifically the number of cows.

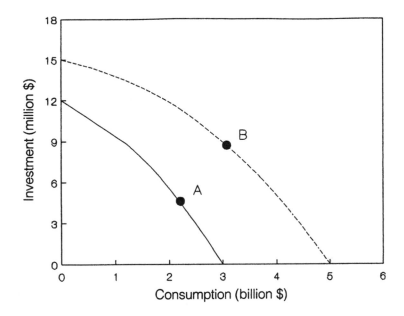

Figure 2.2. Investment vs. consumption in two countries. Solid line is the production-possibility for country A, broken line is the production-possibility frontier for country B.

As we have shown above, future production can be increased by current investment. However, during the last hundred years, technological invention has been more important than mere thrift. The main factors that account for the spectacular increase in production in the modern age have been alternative power sources, machines, standardized parts, breakdown of complex processes into repetitive operations, and specialization.

2.3 Demand and supply curves, equilibrium and competition

In the previous section we considered production-possibility curves. These curves describe what happens to the production of different goods or services as resources are directed to or away from their production. Production of various goods or services will ultimately be determined by equilibrium of supply and demand. We will deal first with the effect of demand on production, as illustrated graphically by "demand curves." A hypothetical demand curve for milk is shown in Figure 2.3. In this curve the price of milk is plotted as a function of the supply of milk. Note that this price is the price that the market is willing to pay for milk - not the cost of production.

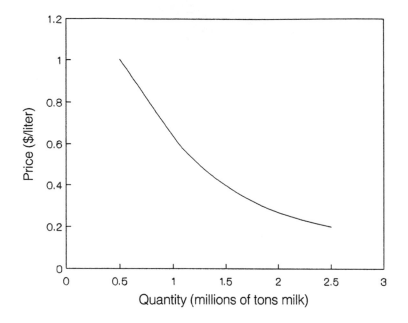

Figure 2.3. Demand curve for milk in a hypothetical country.

Demand curves typically have a negative slope and are convex. The negative slope can be explained as follows: Most people will be willing to pay for a given commodity only if its price is below a certain level. However, in most cases, there will be a small minority who desire the commodity enough to pay a higher price than the rest of the population are willing to do. At limited supply, this minority will determine the demand price of the commodity. As the supply increases, the demand price will be determined by those potential buyers not included in the original group of buyers, who are willing to pay the next highest price. Demand curves are convex because for most goods or services, no matter how great the supply, as long as it is not unlimited, there will always be some minimal price that people will be willing to pay. Thus as a commodity becomes very abundant its price will go down less than its supply will increase.

Conversely, as a commodity becomes very scarce, its price will go up much faster than its supply decreases.

We will now deal with the complementary situation for supply. The hypothetical supply curve for milk production is shown in Figure 2.4, superimposed on the demand curve of Figure 2.3. Note that although the scales for both curves are the same, and both curves are convex, the slope of this curve is positive. Thus as more milk is produced, the cost of production of each unit increases. This of course corresponds to the law of diminishing returns explained in the previous section. At limited production, only those resources which are most efficient at production will be utilized, and the price will be low. As production increases, less efficient resources will be directed to production, and the price of production will increase.

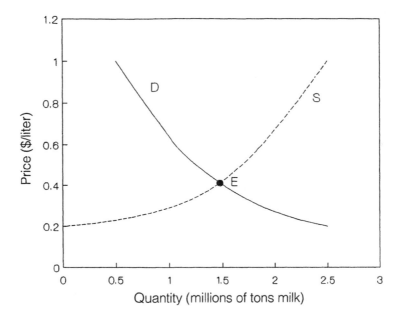

Figure 2.4. Supply and demand curves for milk in a hypothetical country. D is the demand curve, and S is the supply curve. E is the point of equilibrium between supply and demand.

Unlike demand curves, supply curves generally have a positive y-intercept, which means that there is a minimum price of production, no matter how little of a commodity is produced. Supply curves tend to be convex, because overcoming each limiting factor is more difficult than the next. Eventually a situation is reached of maximum possible production of a specific commodity, regardless of cost. In this case the slope of the supply curve tends to infinity.

   In Figure 2.4 there is a point of intersection between the demand and supply curves. With production at the level of this point, the price of production is equal to price of demand. This is the point of equilibrium between supply and demand, and barring outside interference, this will be the market price of the commodity in question. The consequences of prices other than the equilibrium price can be inferred from Figure 2.4. If, for example, the price of milk is above the equilibrium point, then demand for this commodity will be less than the supply. This will cause a decrease in production, which lowers the price of production. The magnitude of the demand at the new, lower, price of production will be greater than the previous level. If the demand quantity is still less than the quantity supplied, then production will decrease further until the equilibrium point is reached. Conversely, if the level of production is below the equilibrium point, then demand for this commodity is greater than supply. This will cause increased production, which will increase the cost of production, but decrease the quantity of the demand, again until equilibrium is achieved.

   As seen above in section 2.2 it is possible through investment to change the quantity of goods and services that a society can produce. If production becomes more efficient through, for example, mechanization, or genetic improvement of agricultural organisms, then it will be possible to produce more at the same cost of production. In other words the supply curve will shift to the right. This situation is illustrated in Figure 2.5. Assuming that the demand curve has remained constant, a new equilibrium point is reached with greater production at a lower price per unit. (Note that this case is different from the situation discussed above where supply was changed, but the supply curve remained constant.) Similar to the supply curve, the demand curve also tends to shift over time. Most advanced societies have in recent history become more affluent. This causes the demand curve for most commodities and services to shift to the right. Thus, at a given price demand will increase. As in the case of a shift in the supply curve, this will also cause a shift in the equilibrium point to the right, but unlike the previous case, the equilibrium price will be higher.

   The examples described until now in this section deal with a situation of "perfect" competition. Near perfect competition will occur only if there are numerous buyers and sellers for each commodity, and a free transfer of information about the prices and quantities for sale. This is hardly ever the case for agricultural products. In later sections we will describe the major factors that interfere with perfect competition and explain their effect on the price and quantity of goods actually produced.

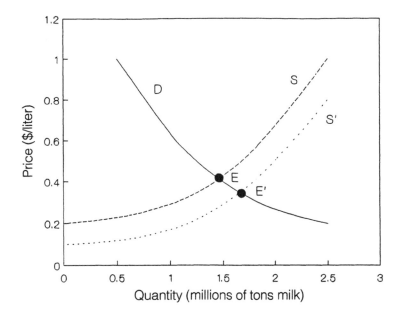

Figure 2.5. A shift in the supply curve for milk. S is the original supply curve. S′ is the new supply curve. E is the original equilibrium point. E′ is the new equilibrium point.

2.4 Elasticity of demand and supply

As shown above, as the price of a commodity increases, the demand will decrease. Therefore the slopes of demand curves are generally negative. A highly negative slope will mean that a large change in the price will affect the quantity of the demand only slightly. Conversely, a nearly horizontal slope will mean that a small change in the price will have a large effect on the size of the demand. The relative change of price and quantity along the demand curve is measured quantitatively by the "elasticity" of the demand. Mathematically elasticity is computed by the following equation:

$$E_d = - \frac{P_r dQ}{Q dP_r} \qquad [2.1]$$

where E_d is the elasticity of the demand, P_r and Q are the price and quantity at a given point on the demand curve, and dQ/dP_r is the derivative of Q with respect to P_r, which is equal to the inverse of the slope at the point P_r, Q. When E_d is greater than unity, then the demand is termed "elastic"; and when E_d is less

less than unity, then the demand is termed "inelastic". We will now define total revenue, or returns (R) as the product of P_r and Q. The change in revenue as a function of P_r and Q can be computed as follows:

$$\frac{dR}{d(P_rQ)} = \frac{P_rdQ}{dR} + \frac{QdP_r}{dR} \qquad [2.2]$$

Maximum revenue will be obtained when its differential is set to zero. In this case the following equations can be derived:

$$-\frac{QdP_r}{dR} = \frac{P_rdQ}{dR} = \frac{P_rdQdP_r}{dP_rdR} \qquad [2.3]$$

Dividing both sides by dP_r/dR, and rearranging gives:

$$1 = -\frac{P_rdQ}{QdP_r} = E_d \qquad [2.4]$$

That is, revenue will be maximum when E_d is equal to unity. A reduction in P_r will increase total revenue if the demand is elastic, and decrease total revenue if the demand is inelastic.

In most cases the demand curve will be convex, dQ/dP_r will vary along the curve, and elasticity will be different at different levels of P_r and Q. However, even if the demand curve is a straight line, elasticity will still be different at different points along the curve because the ratio P_r/Q will be different. Thus for a linear demand curve with a negative slope, elasticity will increase with increase in P_r and decrease in Q. This is intuitively obvious. If the quantity of a commodity is very large, milk, for example, a slight absolute change in its quantity will affect demand only slightly. Conversely, if the quantity of a commodity is small, for example diamonds, a small change in its quantity will have a large effect on demand. This point is critical in agriculture. As agricultural commodities become more prevalent, due to breeding or other means, demand tends to become less elastic, and although prices continue to decline, they decline by increasingly smaller margins.

Elasticity of supply, E_s, can be defined in a similar manner to elasticity of demand:

$$E_s = \frac{P_rdQ}{QdP_r} \qquad [2.5]$$

Note that this equation is the same as [2.1], except for the minus sign, and the fact that dQ and dP$_r$ are now defined relative to the supply curve, rather than the demand curve. Elasticity of demand is more useful than elasticity of supply, because only the former quantity can be used as an indication of the change in total revenue.

An important principle with respect to supply curves is that elasticity of supply tends to increase over the long term. This is because supply curves are different from a short- or long-term perspective. This can be illustrated by the example of milk production. First, consider the momentary situation. The dairy industry at any point in time is geared to produce a curtain quantity of milk. If the price of milk were to double tomorrow, the quantity of milk produced would hardly increase, since there is no way to significantly increase production so rapidly. Thus the "momentary supply curve" is very inelastic. If the price remains high over a period of several weeks or months, then production could be increased somewhat, by additional feed, more frequent milking, and less culling of cows. Of course all these changes will increase the price of production so that the "short-term supply curve" is moderately elastic. Over a period of several years, milk production could be doubled by increasing the number of cows in production. Although this change would take years, the new price of production would be only slightly higher than the original price. Thus "long-term supply curves" tend to be very elastic. In the same way that we defined momentary, short-term, and long-term supply curves, we can define momentary, short-term, and long-term equilibrium.

2.5   Application of supply and demand: "imperfect competition", tax, and price control

The previous section dealt with the case of "perfect competition", where the effect of each individual producer or consumer is infinitesimal on the total supply or demand. In the real world this is seldom the case. Although the principles described above are still true, it is necessary to take account of various factors that can affect supply, demand, and the market price of goods and services. These can be grouped into factors that are external to the market, and factors that are part of the market. We will consider first external factors, the most important of which is government.

The definition of a "government" is a "tax-levying body," and all governments affect price by levying taxes. There are different ways that governments tax, and their effects are different. We will describe only the situation for a unit tax on a commodity. This case is illustrated in Figure 2.6. The tax does not affect the demand curve, but shifts the supply curve to the left. A new equilibrium point is set at a higher price, and at a lower quantity. If the demand is elastic, then the new equilibrium price will be only slightly higher than the previous point, and most of the tax will be born by the producer, who will see both the quantity and the price he receives reduced. If the demand is inelastic, then most of the tax burden will fall on the consumer. The revenue that the producer receives, and the quantity sold, will decrease only slightly.

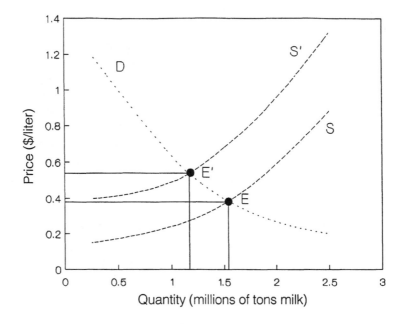

Figure 2.6. Effect of a tax on supply and demand. D is the demand curve, and S is the original supply curve. S' is the supply curve after imposition of the tax. E is the original equilibrium point. E' is the equilibrium point after imposition of the tax.

In addition to taxation, governments can try to directly affect the quantity or price of goods sold by price ceilings, price supports, or rationing, the most common method being price ceilings. The effect of a price ceiling on supply and demand curves is illustrated in Figure 2.7. Without government intervention, production and price would be at point E. By imposing a price ceiling, the price is set at some lower level, p'. At this price, the quantity demanded is Q_d, while the quantity supplied is Q_s. Since Q_d is greater than Q_s, society must have some mechanism for allocating the commodity in question. It will generally be either be on a first-come-first-served basis, or by rationing. Clearly neither of these alternatives is desirable. Another byproduct of price ceilings is black markets, where these commodities are sold illegally at prices above the ceiling. Therefore it is not surprising that although price ceilings have been imposed many times by many different governments, they have rarely achieved the declared goal of providing the consumers with the desired goods at reasonable prices.

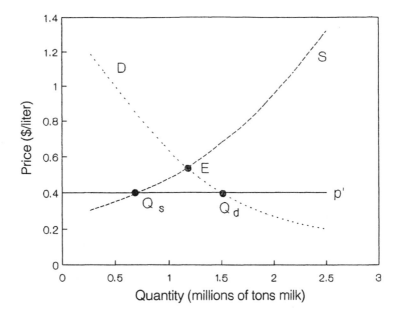

Figure 2.7. Effect of a price ceiling on supply and demand. D is the demand curve, and S is the supply curve. Without government intervention, production and price would be at point E. By imposing a price ceiling, the price is set at p'. At this price, the quantity demanded is Q_d, while the quantity supplied is Q_s.

Price supports have been generally imposed to help producers, and have been more successful in obtaining these goals in the short- and medium-run. As described in the previous section, short-term supply curves tend to be very inelastic, and this is especially true of agricultural commodities. Once the farmer has produced his crop, he has virtually no choice but to sell it at the market price. If the market price is below the price of production, and demand is inelastic, this can lead to a drastic reduction in future production.

This problem is especially severe in animal production. A farmer can decide to plant three times as much wheat next year, but there is no way to increase milk production by a similar factor in so short a period of time. The justifications for price supports are therefore: 1) to assure adequate supplies of desired commodities, and 2) to compensate producers for radical changes in market prices due to factors beyond their control. An example of price support is illustrated in Figure 2.8. In this case the demand curve is shifted so that of decreasing with increased production, it stays parallel to the x-axis. This is accomplished by the government buying any quantity offered at the price P_d, which is above the equilibrium price. Thus the equilibrium price will be P_d, which is higher than the equilibrium price in the absence of a price support. Although price supports do generally solve the problems they are meant to address, they raise both government expenditures and the price consumers must pay.

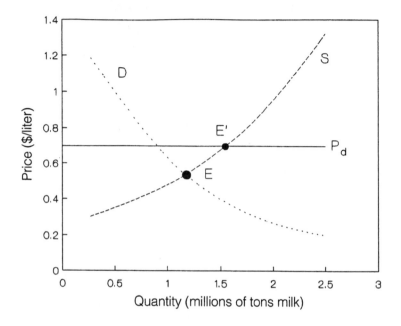

Figure 2.8. Effect of a price support on supply and demand. D is the demand curve, and S is the supply curve. The government will buy any quantity offered at the price P_d, which is above the equilibrium price at E. E' is the new equilibrium point, with price P_d, which is higher than the equilibrium price in the absence of a price support.

Other than governments, large producers or consumers can also affect supply and demand curves. We will consider only the effects of large producers, and will deal with three common situations: monopolies, oligopolies, and cartels. In a monopoly, there is only one producer of a given good or service. This of course does not affect the demand curve, but does render the supply curve immaterial. The goal of the monopoly will be to maximize profit. Since by definition, profit is minimal at the equilibrium price, the monopoly will choose to produce less than the equilibrium quantity, because at this quantity, the demand price will be higher. Again defining the demand price and quantity as P_{rd} and Q_d, and the equilibrium price as P_{re}, the monopoly will try to fix production so as to maximize $Q_d(P_{rd} - P_{re})$.

In an oligopoly there are only a few large producers of a given commodity. Thus, contrary to the situation in "perfect competition" a single producer can affect the market by a change in the amount of his production, or the price he charges. Generally in an oligopoly, a single producer can dramatically increase his market share by charging slightly less than his competitors. Since two can play this game, the producers in an oligopoly often try to fix price among themselves. This was the situation prevalent in the airline industry until the 1970's. Of course the rewards for "cheating" are great, and in the absence of outside intervention, this will most likely occur.

In a cartel the producers agree to fix the price, similar to in a monopoly. This can only work if the producers also decide to limit production. Again the rewards for "cheating", either by increasing production above the agreed quota, or selling below the agreed price are great, as has become apparent to the members of OPEC. Thus cartels tend to be inherently unstable, unless government control is imposed.

2.6   Increased efficiency of production, and the dynamic cobweb; converging, diverging, and persistent oscillations

Dickerson (1970) maintains that the primary goal of genetic improvement is to increase the efficiency of production, i.e., to produce the same amount at a lower price, or produce more at the same price. This will result in a shift of the supply curve to the right. Other things being equal, a new equilibrium point will be achieved at a lower price and at a higher quantity of production. However, over the long-term, societies have tended to become more affluent. This has caused demand curves also to shift to the right. The effect of the long-term shifting of both supply and demand curves is illustrated in Figure 2.9. If the original and new equilibrium points are compared, it is clear that, although the quantity produced has increased significantly, the equilibrium price has hardly changed. If the supply curve has shifted more than the demand curve, then there will be a slight decline in the equilibrium price. This answers a commonly asked question, "If production is so much more efficient now, why don't prices go down?" From this figure, it is clear that efficiency and price do not necessary go hand-in-hand.

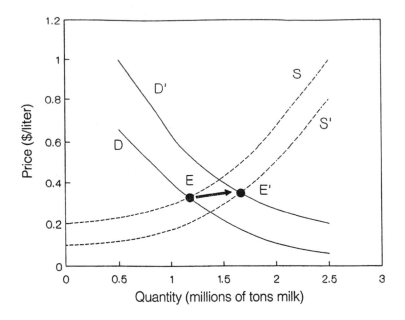

Figure 2.9. Simultaneous shift of the supply and demand curves. D is the original demand curve, S is the original supply curve, and E is the equilibrium point. D′ is the new demand curve, S′ is the new supply curve, and E′ is the new equilibrium point.

What happens when equilibrium of price and production is disturbed? Consider the situation illustrated in Figure 2.10a. Originally production is at level Q_1, which is below the level of equilibrium production Q_e. At this level of production, the market price will be determined by the demand curve, and will be equal to P_{r1}. Note that this price is above the equilibrium price, P_{re}. At this price, production will increase to Q_2, which is the quantity of production of the supply curve for the price P_{r1}. At the new level of production, price will again be set by the demand curve, but this time the market price, P_{r2}, will be below the equilibrium price. It can readily be seen that by decreasing oscillations, production and price will tend toward the equilibrium point.

Not all markets will tend toward equilibrium. In Figure 2.10a, the demand curve was more elastic than the supply curve. Figure 2.10b illustrates an example where the supply curve is more elastic than the demand curve. In this case, each oscillation will move the market farther from equilibrium. Figure 2.10c illustrates the situation where the two curves are of equal elasticity. In this case, price and production will continue to oscillate at a constant frequency. This helps explain why despite the laws of supply and demand, markets are often far from the equilibrium point.

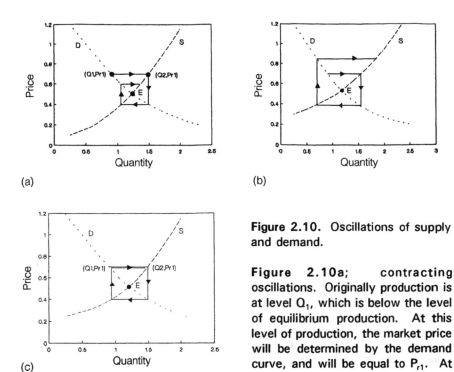

(a)   (b)

(c)

Figure 2.10. Oscillations of supply and demand.

Figure 2.10a; contracting oscillations. Originally production is at level Q_1, which is below the level of equilibrium production. At this level of production, the market price will be determined by the demand curve, and will be equal to P_{r1}. At this price, production will increase to Q_2. Eventually production and price will stabilize at the equilibrium point, E'.

Figure 2.10b; expanding oscillations. The supply curve is more elastic than the demand curve.

Figure 2.10c; constant oscillations. The supply and demand curves have equal elasticity. Price and production will continue to oscillate at a constant frequency.

2.7 Economics of agriculture, long-run decline, short-run instability, and government aid

We will now apply the principles described above to the specific case of agriculture in the developed countries. Over the last two hundred years, the efficiency of agriculture has increased dramatically. This can be illustrated first by the diminishing percent of the population engaged in agriculture. If at the time of the American Revolution, over two thirds of the population of the US was engaged in agriculture, in 1988 only about 5% of the work force was so employed, and the percentage is still declining. Similar trends have occurred in other developed countries, and the causes for increased efficiency have been both agritechnical and genetic.   Not only have more people been fed with proportionately less labor, they have also been fed a richer diet.   Food production from animal sources requires greater inputs in energy than food production from plants.   Nevertheless, consumption of animal products as opposed to plants, and consumption of fat and protein as opposed to carbohydrates has increased.

What has happened to farm income?   As described in Section 2.6, both supply and demand curves tend to shift to the right over time.   Although this results in greater production at equilibrium, the effect on prices is unpredictable. The historic situation in agriculture is illustrated in Figure 2.11.   Supply and demand curves are drawn for two points in time several decades apart.   Both curves shift to the right over time, the supply curve because of increased efficiency of production, and the demand curve because of increased population and affluence.   As an example of the effect of increased affluence on agricultural consumption, it can be noted that from 1967 to 1987, per capita milk butterfat consumption in Israel increased from 4.1 to 6.0 kg (Dror, 1988). However, the change was significantly greater for the supply curve.   Therefore the equilibrium prices of agricultural commodities, adjusted for inflation, tend to decline.

Another economic disadvantage of agriculture, is that farm incomes have tended to fluctuate more than other incomes, despite the fact that farm production is much more stable than non-farm production. Although it is generally thought that agricultural production is at the mercy of the weather, in fact fluctuations in production due to weather conditions are quite minor. Why then do food prices fluctuate so markedly over the short-term? The reason is that both the short-term supply and demand curves are very inelastic.   Thus slight shifts in production will have large effects on the price of agricultural commodities.   It is therefore not surprising that farmers have traditionally looked to government for help. Currently most developed countries have large governmental programs to aid agriculture.

Although this book is about economic aspects of animal breeding, economics cannot be divorced from politics, and politics has clearly played a significant part in agriculture. We will now summarize the major methods governments use to aid agriculture, and their effects on price and production.   First many governments support agricultural research. In Section 2.2 we explained that a society must choose between investment and current consumption. The same is true of a company or an enterprise, and many high-technology industries invest a considerable fraction of total expenditures in research and development.   In agriculture the cost of research, to a large extent, is born by governments, and individual farmers are able to benefit without having to divert their own resources in this direction. However, it should now be clear from the discussion of the dynamic cobweb in the previous section, and Figure 2.11, that although individual farmers, or farmers of a given country may benefit from agricultural research over the short-term, farmers do not gain much in the long-term.   The effect of increased productivity on agriculture has been termed the "progress-surplus-bankruptcy cycle" (Moav, 1973), and will be discussed in more detail in Chapter 4.

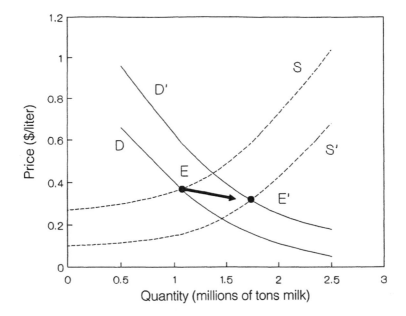

Figure 2.11. Traditional downward shift of farm prices. Supply and demand curves are drawn for two points in time several decades apart. D is the original demand curve, S is the original supply curve, and E is the equilibrium point. D' is the new demand curve, S' is the new supply curve, and E' is the new equilibrium point. Both curves shift to the right over time, but the equilibrium price at E' is lower than at E.

Other methods of governmental aid have chiefly been directed either to increase demand or to reduce production. Governments have employed several methods to increase demand. We have already discussed the effect of price supports, which have been employed extensively in the US. In addition to merely buying and storing excess production, governments of developed countries have sold agricultural products overseas or to needy local consumers at reduced prices. Thus two birds are killed with one stone, in addition to the needs of farmers, other governmental needs, such as foreign aid, or internal relief are met. The problem with these programs is that over time they tend to increase in size and cost. If prices are kept at a constant level by the government, while productivity increases, the differential between market demand and production will increase, causing the government to buy ever larger surpluses.

Governments have tried to limit production by either direct payments to farmers, or by quotas on production. Both of these methods shift the supply curve to the left, and thus compensate for the effect of increased efficiency. The former method, although extensively employed by the US government until 1970, has negative moral complications. "Why should rich farmers be paid not to plant while people are starving?" has been a common refrain. The problem with production quotas is the same problem discussed above with respect to cartels and oligopoly. As long as other producers are abiding by their quotas, there is a strong incentive for the individual producer to cheat. Thus methods are necessary to enforce quotas, with commensurate economic and moral costs.

The final method is direct governmental payments to farmers in distress. This method has little undesirable moral baggage, but is hardly a solution to the long-term problem of excess production. We will return to the question of the role of government in agricultural in Chapter 4.

2.8 Marginal utility and marginal cost, fixed and variable costs, and long-run break-even conditions

Until now in our discussion of supply and demand curves, we have not considered the factors, other than efficiency of production, that determine the relationship between price and production. We will now attempt to explain the underlying economic principles that determine these relationships. We will begin with the demand curve and the concept of "utility".

Consumers are willing to pay for goods and services because they supply some satisfaction or need, which in economic terms is called "utility". It is a general principle that for any particular commodity, although the total utility increases with each unit acquired, the increase in utility diminishes. A single drop of water is important for a thirsty man, but of no practical importance in filling a swimming pool. This leads us to the "Law of diminishing utility." This principle is parallel to the law of diminishing returns, but refers to demand, rather than production.

Figure 2.12 illustrates this principle for the case of butter. Total and marginal utility, defined as the derivative of total utility, are plotted as a function of quantity. It can be seen that total utility increases, while marginal utility decreases. If this is the case, one can then ask how much butter, or how much of any good, will a consumer buy? Clearly this question is related to the price of different goods. For example, one receives more utility from an automobile than a kg of butter, but then the automobile costs more. Thus any consumer will try to achieve maximum utility for the amount of money spent. This will be accomplished only when the marginal utility per unit price of each good purchased are equal.

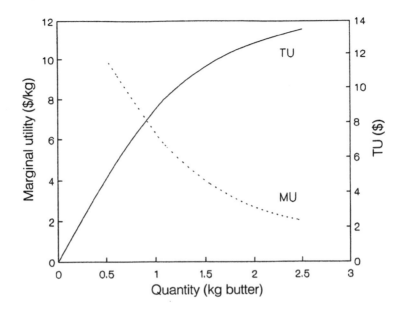

Figure 2.12. Marginal and total utility. TU is the total utility curve, and MU is the marginal utility curve.

The principle is called the "Law of marginal utility" and can be described by the following equation:

$$MU_e = MU_1/P_{r1} = MU_2/P_{r2} = MU_i/P_{ri} \qquad [2.6]$$

Where MU_e is the equilibrium marginal utility, MU_1 and MU_2 are the marginal utilities for goods 1 and 2 at prices P_{r1} and P_{r2}, respectively, and MU_i is the marginal utility for the ith good at price P_{ri}. That is the price of any good will be equal to its marginal utility divided by the equilibrium marginal utility.

From this discussion it is clear that the demand curve will be determined by the ratio of marginal utility to price for each consumer. For supply curves, the determining factor will be the marginal cost of production. The cost of milk production in a dairy herd as a function of total production is plotted in Figure 2.13. Also plotted is the marginal cost, which is the derivative of total cost. If we first study the curve for total production, we note that the curve does not start at the origin. In other words a sizable initial investment is necessary before any milk is produced. Once this initial investment is made, total cost increases as a function of total production, but the slope is not constant, as illustrated by the plot of marginal cost. At low production, the slope is negative, while at high production, the slope is positive. That is the general case for marginal utility, and can be explained as follows. If the farmer has facilities to handle 100 cows, but has only 50, he will still have to cover all the costs of building and equipment. Thus in this case he can increase production with only a moderate increase in his total costs. However, once he reaches the level of 100 cows, production can be increased only by better management or more frequent milking. Of course there is a limit to how much these factors can increase production, and at that point, additional expenditures can increase production only slightly.

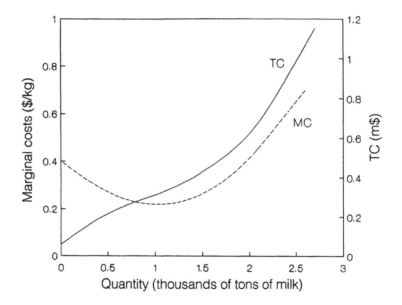

Figure 2.13. Marginal and total costs of milk production. TC is the total costs curve, and MC is the marginal costs curve.

How much will the enterprise produce at equilibrium? If the market price is above the marginal cost of production, it pays for the firm to produce more, because in this case the return on the additional production will be greater than the cost of the additional production. However, if the marginal cost is above the market price, then the firm is losing money on each additional unit produced, and should therefore reduce production. Thus at equilibrium, price will be equal to marginal cost. This does not mean that the total cost of production is not important. The practical relationship between total and marginal cost is illustrated in Figure 2.14. Three possible demand curves are drawn together with the marginal cost curve. At each level of demand, the production of the enterprise will be set so that P_r = MC. At D_1 total revenue is above total cost of production, and there is a profit. If in this case production is decreased, MC will also be lower, but profit, equal to total revenue - total cost of production will also decline. At demand D_2, total revenue is equal to the total cost of production when P_r = MU. In this case, profit is equal to zero, and there is no incentive for new producers to enter the market. This is the case of market equilibrium between supply and demand discussed above. If demand falls to the level of D_3, the firm will still produce up to the level where P_r = MU, but total revenue will be less than the total cost of production. This situation is clearly unstable over the long-term. If demand falls even lower, then total revenue will no longer cover the variable costs of production, that is those costs that depend on the amount produced. In this case no production is economically justified. This is the shutdown point marked on the MC curve.

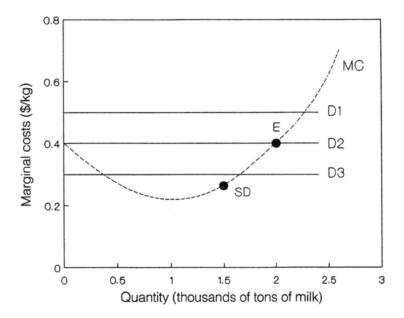

Figure 2.14. Marginal costs with three levels of demand. MC is marginal cost curve. At demand D_1 total revenue is above total cost of production, and there is a profit. At demand D_2, total revenue is equal to the total cost of production. At demand D_3, total revenue is less than the total cost of production. E is the equilibrium point, and SD is the shutdown point marked on the MC curve.

Figure 2.15 summarizes the relationship between supply and demand, total utility and total cost, and marginal utility and marginal cost. Total utility, total cost and marginal cost increase with increased production, while marginal utility decreases. The curves for marginal utility and marginal cost are in fact the demand and supply curves, and their point of intersection is the equilibrium point between supply and demand.

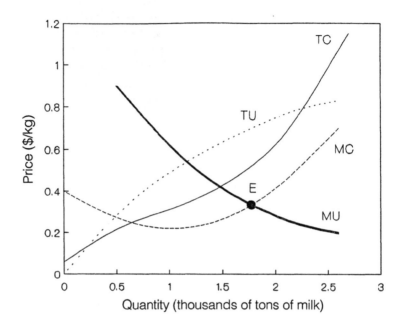

Figure 2.15. Marginal and total costs and utility. TC is total costs, TU is total utility, MC is marginal costs, and MU is marginal utility. E is the equilibrium point at the intersection of marginal costs and utility.

We have seen above that it is useful to divide costs into two categories, fixed and variable. Total costs are defined as the lowest aggregate expense needed to produce each level of production. Fixed costs are all costs independent of the amount produced, while variable costs are costs dependent on the level of production. Thus for a dairy enterprise, buildings and land are fixed costs, while feed and labor are variable costs. Average total, fixed and variable costs (AC, AFC, and AVC), are defined respectively as total, fixed and variable costs divided by total production.

The relationships between average costs and marginal cost are illustrated in Figure 2.16. As explained above, the curve for marginal cost is convex. The curve for average fixed costs will have a negative slope, and tend asymptotically to zero. This is because as production increases, fixed costs by definition remain the same. Since average fixed costs is the ratio between a constant and total production which increases, the average fixed costs decline. Average variable costs will also tend to be convex, but will have a minimum value. This is because a considerable amount of feed and labor will be necessary to produce the first kg of milk. After that average variable costs will decline because additional production will require only a slight increase in the variable costs. However, as production increases, average variable cost will start to increase, due to the law of diminishing returns. Average total costs are the sum of average variable and fixed costs, and will also be convex with a minimum. Note that the curve of marginal costs intersects both the curves of average variable and total costs at their minimums. This is because average variable and total costs will be declining as long as they are greater than marginal costs, and increasing if they are less than the marginal cost.

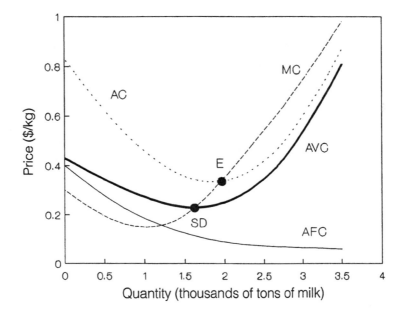

Figure 2.16. Marginal, average fixed and variable costs, and total costs. MC is marginal costs, AC is average costs, AVC is average variable costs, AFC is average fixed costs. E is the equilibrium point, and SD is the shutdown point.

The long-term equilibrium point, which is also the break-even point, is the point where MC, AC, and the demand curve intersect. If demand is below this point then production will decrease, but total revenue will still be less than total costs, and the producer will be losing money. If demand is above this point, then production will be at the intersection of MC and demand, which will be greater than AC. In this case the producer will have a profit, and other producers are likely to begin or increase production. The intersection of MC and AVC is the short-run shutdown point. If demand falls below this level production will cease completely, because the revenue will no longer cover the variable cost of production.

It should now be clear that although all producers will try to maximize their profit, profits will nevertheless remain minimal over the long-term due to competition. Thus it is the consumer who tends to gain from increased efficiency of production over the long-term. The reader may ask how can there be a stable equilibrium without any profit for the producer? This consideration is usually handled by assuming that a "reasonable" profit is part of the costs of production.

2.9 Interest, discount rates, inflation and profit horizon

We have discussed above that a society must choose between current consumption and investment. Likewise an enterprise must choose between current production and investment in research and development. Why is investment desirable? Because there are indirect processes, which take time to get started, but are more productive than direct processes in the long-term. Breeding programs are a good example of this principle. All breeding programs cost money, but yield returns only in the future. The main cost elements of breeding programs are: data collection, keeping of non-productive animals for future breeding, test matings, and statistical analysis. Rather than keep live animals it is possible to keep breeding stock in the form of frozen semen or embryos. This saves the costs involved in keeping non-productive animals, but entails additional costs for the production and preservation of the "seed." Recent biotechnology advances have raised the possibility of new techniques that could increase the rate of genetic advance, but would also increase the costs of breeding programs. These techniques will be considered in detail in Chapter 12.

Thus capital has a net productivity which can be measured as a "real" interest rate. (By "real" interest rate, we mean interest rate corrected for inflation, which will be discussed below.) If investment is so good why do societies not invest more? As described above, the amount a consumer or a society is willing to spend on each commodity is determined by marginal utility. Thus a society will invest to the level that the marginal utility of investment is equal to the marginal utility derived from other commodities. This is determined by the concept of "minimum attractive rate of return" (MARR). This concept can be explained as follows: money now is worth more than money received in the future. Therefore in order to invest money now, future return should be greater than the amount invested. The difference per annum is called the minimum attractive rate of return. Using this principle we can define the discounted value of a current investment as follows:

$$DV = \frac{N}{(1 + d_i)^t}$$ [2.7]

where DV is the discounted value of N current dollars, d_i is the minimum rate of return (MARR), and t is the time in years between investment and realization of the return. Thus for N = \$100 and d_i = 10%, an investment that matures in one year will be attractive only if the return is at least \$110. Another way of considering the problem is to say that those investments which will realize the highest rate of interest will be chosen first. All investments with interest rates greater than d_i will be chosen, but not those with interest rates below d_i.

There are many ways to invest money, and to receive returns, generally both will occur over extended periods of time. A common situation is when a certain sum is invested now, and a constant sum is paid back each year in perpetuity. Assuming that N dollars are invested, and that MARR = d_i, the minimum acceptable annual return, V, can be computed as follows:

$$Nd_i = V$$ [2.8]

Thus if in the previous example, N = \$100, and d_i = 10%, V = \$10. That is, if my MARR is 10%, an investment of \$100 now under the conditions just described is justified only if annual return is at least \$10 each year. Investment situations in breeding programs are generally more complicated than the two situations described, and will be described in more detail in Chapter 8.

Nearly all modern societies suffer from some level of inflation, thus nominal interest rates will be higher than "real" interest rates. Assume that \$100 today buys 400 kg milk. \$100 next year with inflation will be worth less a) because I would rather have the milk now, and b) \$100 will buy less milk in a year from now. With an inflation rate of d_t, the discounted value of current investment can be computed as follows:

$$DV = \frac{N}{[(1 + d_q)(1 + d_t)]^t}$$ [2.9]

Where d_q is the "real" interest rate, corrected for inflation. Although nominal interest rates and inflation rates have varied dramatically in many societies, long-term real interest rates have remained remarkably constant in the range of 3 to 5 percent (Smith, 1978). For animal breeding the nominal interest rate will have no significance, since future investment is returned in commodities - not in

currency.   Thus investments in breeding programs have continued even in
societies with very high inflation rates.

Two important differences between breeding programs and nearly all other
investment are that all gains in breeding programs are cumulative, and for
perpetuity. This important point can be illustrated by the following example.
Recent studies have shown that injection of cattle with bovine somatotropin
(growth hormone) will increase milk production by 10 to 20 percent (Leitch,
Burnside, and McBride, 1990). Clearly this is quite impressive, especially if the
cost of the treatment is low. However, the improvement will be realized only as
long as cows continue to be injected. If, on the other hand, milk production is
increased by breeding, this differential will be maintained for eternity without
further treatment. Furthermore, any additional genetic improvement will yield
a cumulative result to previous breeding.

Since breeding is generally a long-term proposition, and gains are perpetual
and cumulative, the question of how far into the future one should consider
returns is important. Beyond considerations of a minimum attractive rate of
return, it can be argued that returns too far in the future have a present value of
zero. Thus long-term breeding programs are generally considered in terms of
a "profit horizon". That is all returns occurring after the profit horizon, are
considered to have no value. Breeding programs are generally analyzed in terms
of a ten to twenty year profit horizon.

2.10 Summary

This chapter described the main economic concepts applicable to animal
breeding. We showed the necessity of choosing between alternative economic
goods, and explained the law of diminishing returns. Graphic display of demand
and supply was used to illustrate the conditions for equilibrium, how changes in
price and production were affected by elasticity of supply and demand, and the
effects of government intervention and increased efficiency of production. The
concepts of utility, and marginal, fixed and variable costs were introduced to
explain demand and supply curves. We showed that the amount of investment
will be determined by the minimum attractive rate of return, and the profit
horizon. The recent economic trends in agriculture in developed countries were
summarized.

Chapter Three

Principles of Matrix Algebra and Selection Index

3.1 Introduction

Hazel in 1943 formulated the principles of economic selection index. He asked the following question: Assume that there are n traits for which breeding values can be estimated, and m traits with economic values. Assume further that the economic values of the m traits are linear functions of the trait values. Some, but not all of the traits included in n may be included in m, and *vice versa*. What linear index of the n measured traits should be used to select individuals so as to maximize genetic progress on the economic scale? Hazel formulated his principles without the benefit of matrix algebra, which was not then used commonly by geneticists. It later became clear that the principles of selection index could be expressed more succinctly and generally through this important mathematical tool. Since matrix algebra will also be of major importance for much of the remainder of this book, Section 3.2 reviews the principles of matrix algebra, for those readers who are unfamiliar with the topic. *Matrix Algebra Useful in Statistics* (Searle, 1982) is recommended for a more extensive study of this topic. In Section 3.3 we will discuss the concepts of genetic and environmental correlations, and demonstrate how selection on one trait will lead to correlated responses on other traits. In the remaining sections of this chapter we will derive the selection index equations, describe its properties, and give some examples of the practical use of selection index.

Although we have no desire to belittle Hazel's accomplishment, the limitations of the method should also be noted. Especially since there has been a tendency in the literature to assume that selection index is the complete answer to the topic of this book. It should first be noted that selection index, as formulated by Hazel, only solves the case where the economic values are linear functions of the trait values. In Part 2, it will be shown that this is generally not the case. Second, selection index assumes that the economic values are known *a priori*. Not only is this seldom true, economic values tend to change over time and place. Finally, selection index only provides relative weights between traits, it does not answer the question of whether breeding is economically justified.

3.2 Principles of matrix algebra

We will first introduce the concept of a vector, which is essentially just a series of numbers. A vector can be displayed either as a horizontal or vertical series of numbers. A horizontal series will be called a "row vector", and a vertical series will be called a "column vector". A matrix is a two dimensional array of numbers, and can be represented as follows:

$$\begin{bmatrix} 7 & 3 & -5 \\ 8 & 4.2 & 0 \end{bmatrix}$$

This matrix has two rows and three columns. A matrix with only one row can be considered a row vector, while a matrix with only one column can also be termed a column vector. A matrix with only one row and column, i.e. a single number, is termed a "scalar". The individual numbers in a matrix are called the "elements" of the matrix, and are denoted by subscripts. Generally matrices will be denoted by **bold** uppercase letters, vectors by **bold** lowercase letters, and scalars and matrix elements by regular face letters. For example, if the matrix above is denoted as "**A**", a specific element of the matrix can be denoted as a_{ij}, where the first index refers to the row number, and the second index to the column number. Thus $a_{12} = 3$.

A few more useful definitions will now be presented. The transpose of a matrix is a matrix in which the elements of the columns are replaced by the corresponding row elements. Thus the transpose of the previous matrix is:

$$\begin{bmatrix} 7 & 8 \\ 3 & 4.2 \\ -5 & 0 \end{bmatrix}$$

A matrix with equal number of rows and columns is called a "square matrix". The diagonal of a square matrix from the upper left to lower right corners is called the "diagonal" of the matrix. The other diagonal is called the "secondary diagonal". The sum of the elements of the diagonal is called the "trace" of the matrix. A square matrix in which the elements above the main diagonal are a "mirror image" of the elements below the diagonal is called a "symmetrical matrix". The following is an example of a symmetrical matrix:

$$\begin{bmatrix} 7 & 2 & -5 \\ 2 & 3 & 4.2 \\ -5 & 4.2 & 0 \end{bmatrix}$$

The transpose of a symmetrical matrix will be equal to the original matrix.

Mathematical operations can be performed with matrices that parallel the scalar mathematical operations of addition, subtraction, and multiplication. The matrix operation parallel to division is called "matrix inversion". We will briefly describe these operations. In matrix addition each element of one matrix is added to the corresponding element of the second matrix. Similarly, in matrix subtraction the elements of one matrix are subtracted from the corresponding elements of the second matrix. It is possible to perform matrix addition or subtraction only on two matrices with the same number of columns and rows.

Matrix multiplication is slightly more complicated. We will therefore start first with the example of multiplication of a row vector by a column vector, both with the same number of elements. In this case each element of the row vector is multiplied by the corresponding element of the column vector, and these products are summed. The matrix product is then a single number, i.e. a scalar. In the opposite case, that is multiplication of a column vector by a row vector of equal number of elements, the product will be a square matrix. Each element will be the product of the corresponding elements of the row and column vectors. For example, the element for row 2 and column 3 will be the product of element 2 of the row vector, and element 3 of the column vector. For the general case of multiplication of two matrices, the element for the ith row and j^{th} column will be the sum of the products of the elements of the ith row of the first matrix, multiplied by the elements of the jth column of the second matrix. Thus it is possible to multiply two matrices only if the number of columns of the first matrix is equal to the number of rows of the second matrix. The resultant matrix will have as many rows as the first matrix, and as many columns as the second matrix. For example the product of a matrix times a column vector will be a column vector, while the product of a row vector by a matrix will be a row vector. Multiplication in the opposite order, for example a column vector times a matrix, is not defined.

Before discussing the final operation, matrix inversion, we will first define a special matrix called the "identity matrix". This is a square, symmetrical, matrix with ones on the diagonal, and zero for all non-diagonal elements. Identity matrices are denoted by "I". It can readily be shown that multiplication of any matrix by an identity matrix will result in a product equal to the first matrix. The parallel in scalar arithmetic is multiplication by unity.

We will now define the inverse of a matrix through an example. Assume a matrix **A**. The inverse of **A** is the matrix that when multiplied by **A** yields an identity matrix. The inverse of a matrix is generally denoted by the negative first power. Thus algebraically:

$$A*A^{-1} = I \qquad\qquad [3.1]$$

We will use the star to define matrix multiplication. This is of course parallel to scalar division, in which the product of a number and its inverse is equal to unity. Only square matrices have a unique inverse, but not all square matrices can be inverted. A square matrix with a unique inverse is called a "nonsingular matrix", while a matrix without a unique inverse is called a "singular matrix." For singular matrices a matrix called a "generalized inverse" can be computed which has some of the important properties of a "true" inverse.

Unlike matrix addition, subtraction, and multiplication, there are no simple algorithms for matrix inversion. Rather complicated algorithms have been developed that can be used to invert all nonsingular matrices. However the amount of computer time required increases exponentially with the number of row (or columns) in the matrix. Thus even with modern computers, it is quite time consuming to invert very large matrices. However, for certain important matrices short-cut algorithms have been found (Henderson, 1976).

The main use of matrix algebra is to solve systems of linear equations. We will illustrate this with the following example:

$$5x_1 + 3x_2 - 4.2x_3 = 21$$

$$3x_1 - 8x_2 + 5x_3 = 10 \qquad\qquad [3.2]$$

$$8x_1 + 10x_2 - 3x_3 = -5$$

It should generally be possible to solve the following system of three equations for the three unknowns of x_1, x_2 and x_3. Using matrix algebra this system of equations can be written as follows:

$$\begin{bmatrix} 5 & 3 & -4.2 \\ 3 & -8 & 5 \\ 8 & 10 & -3 \end{bmatrix} \begin{bmatrix} x_1 \\ x_2 \\ x_3 \end{bmatrix} = \begin{bmatrix} 21 \\ 10 \\ -5 \end{bmatrix} \qquad\qquad [3.3]$$

Let us call the first matrix the coefficient matrix, and denote it A. The two vectors will be called the "solution vector" and the "right-hand-side vector", and will be denoted x and y, respectively. These equations can then be denoted as follows:

$$A * x = y \qquad [3.4]$$

In order to solve for x we multiply both sides by the inverse of A. Algebraically:

$$A^{-1} * A * x = A^{-1} * y \qquad [3.5]$$

Since a matrix times its inverse is equal to the identity matrix, and the product of any matrix and the identity matrix is the original matrix, we have:

$$x = A^{-1} * y \qquad [3.6]$$

Thus using matrix algebra we are able to solve any system of linear equations for which the inverse of the coefficient matrix can be computed.

Similar to scalars, calculus operations can also be performed on matrices. For example, the differentials of $x'*A$ and $x'*A'*A*x$ with respect to x will be A and $2A'*A*x$, respectively.

We will finish this section with a brief summary of the least squares equations. (From this point on we will delete the star to signify multiplication as is generally done.) Assume that there exists a system of equations such as that given in equation [3.4] for which no solution exists. This can be either because there are more equations than unknowns (in this case A will not be a square matrix), or because there is a linear dependence between the equations. This system of equations can be phrased as follows:

$$y = Ax + e \qquad [3.7]$$

where y is a known vector, A is a known matrix, and x and e are unknown vectors. x is termed the vector of "solutions", and e is termed the vector of "residuals". For example, y can be a vector of measurements made on a sample of animals: weight, height, milk production, etc., with A, a matrix of known "treatments" applied to these animals. x will then be the vector of the effects of these treatments on this sample of animals. We wish to solve for the values of x that minimize the sum of squared residuals. This sum of squares will be equal to the squared difference between Ax and Y. That is we wish to minimize the following quantity:

$$e'e = (y - Ax)'(y - Ax) = x'A'Ax - 2x'A'y + y'y \qquad [3.8]$$

Differentiating with respect to x and setting equal to zero we obtain:

$$2A'Ax = 2A'y \qquad [3.9]$$

and finally:

$$x = (A'A)^{-1}A'y \qquad\qquad [3.10]$$

Equations [3.10] are termed the "normal equations", and are used to derive the least squares solutions for x. If the matrix A is singular, then a generalized inverse of A'A can be substituted for the true inverse. In this case, there is no unique solution to the normal equations, but it is possible to obtain a specific solution, by adding a constraint to the system of equations, for example the constraint that all levels of an effect should sum to a given value. This is achieved by solving the following system of equations:

$$\begin{bmatrix} A'A & c \\ c' & 0 \end{bmatrix} \begin{bmatrix} x \\ \tau \end{bmatrix} = \begin{bmatrix} A'y \\ n \end{bmatrix} \qquad\qquad [3.11]$$

Where c is a vector of 1's, τ is a "Lagrange multiplier" and n is the desired sum of all levels of the effect. Thus, it is necessary to solve the equation c'x = n, in addition to the normal equations. The additional equation in this system "breaks" the dependency in A'A.

3.3 Genetic and phenotypic variance-covariance matrices and correlated response to selection

In the first chapter we showed how response to selection on a single trait will be dependent on the ratio of genetic to phenotypic variance for that trait. We will now discuss the situation of more than a single trait in the context of scalar and matrix algebra. When several traits are measured on each individual, it is possible to compute phenotypic covariances among the traits. A matrix can then be constructed called the phenotypic variance-covariance matrix with trait variances on the diagonal and covariances on the off diagonals. For example, assume that two traits are measured on dairy cows, milk production and fat percent.   Assume further that the phenotypic variances for milk and fat percentage are 2,000,000 kg^2 and 0.12%^{2}, respectively, and that the covariance is -294 kg-%. The phenotypic variance-covariance matrix will then be:

$$\begin{bmatrix} 2,000,000 & -294 \\ -294 & 0.12 \end{bmatrix}$$

Note that a variance-covariance matrix is always a square symmetrical matrix. We can similarly construct the genetic variance-covariance matrix, with the genetic variances on the diagonals and the genetic covariance on the off diagonal. The genetic covariance will be equal to the fraction of the total covariance determined by additive genetic factors. In this example we will assume the following genetic variance-covariance matrix:

$$\begin{bmatrix} 500,000 & -86.6 \\ -86.6 & 0.06 \end{bmatrix}$$

In this example the heritability of milk production is: $500,000/2,000,000 = 0.25$. The heritability of fat percent is: $0.06/0.12 = 0.5$. The phenotypic correlation is computed as the phenotypic covariance divided by the square root of the product of the phenotypic variances, and the genetic correlation is computed as the genetic covariance divided by the square root of the product of the genetic variances. In this case the genetic correlation, r_g is computed as follows:

$$r_g = \frac{-86.6}{\sqrt{(500,000)(0.06)}} \qquad [3.12]$$

If two traits have a non-zero genetic correlation, then selection on one will lead to a genetic change on the other. This change is called the "correlated response", and can be estimated as follows: Assume that we select directly for trait X. The response for this trait will be equal to the difference of mean breeding value of the selected individuals from the population mean. (As previously we will assume that the population mean is equal to zero. If trait Y is genetically correlated to X, then the change in Y will be the regression of the breeding value of Y on X, $b_{(A)YX}$, which can be computed as follows:

$$b_{(A)YX} = \sigma_{AXY}/\sigma^2_{AX} = r_g(\sigma_{AY})/\sigma_{AX} \qquad [3.13]$$

where σ_{AXY} is the genetic covariance between the traits; σ^2_{AX} is the genetic variance for x; and σ_{AY} and σ_{AX} are the genetic standard deviations for x and y, respectively. The correlated response will then be equal to this regression times the direct response on X. Thus:

$$\phi_{Y/X} = b_{(A)YX}\phi_X \qquad [3.14]$$

where $\phi_{Y/X}$ is the correlated response of y to selection on x, and ϕ_X is the direct response of X. From Chapter 1, equation [1.24], we have:

$$\phi_X = ih_X\sigma_{AX} \qquad\qquad\qquad [3.15]$$

Therefore:

$$\phi_{Y/X} = ih_X r_g \sigma_{AY} \qquad\qquad\qquad [3.16]$$

Thus the correlated response can be computed as a function of the selection intensity on trait x, and the variance components of x and y.

3.4 Derivation of the selection index

Although Hazel did not phrase his derivation of selection index in matrix terms, we will do so because it greatly facilitates explanation. We have already shown that, for selection on a single trait, the rate of genetic improvement will be a function of the intensity of selection, the accuracy of the evaluation, and the genetic variance. We have also shown that selection on a single trait can cause a correlated response on other traits. Generally several traits have economic value in a species under selection. How then should selection be performed so as to economically maximize genetic improvement?

We will start by assuming that for each individual there is a vector y, of length m, consisting of the individual's breeding values for traits of economic importance and a vector x of n measured traits to be included in the selection index. Although x and y may include the same traits, this does not have to be the case. Assume further that the "economic values" associated with y are linear functions of the trait values. (Methods to derive the economic values are discussed in Part II.) We can then define a vector a, also of length m, consisting of the economic values of the traits in y. The aggregate economic breeding value, H, can then be computed as a'y. The units y are trait units, and the units of a are monetary units/trait units, for example dollars/kg milk. Thus H is a scalar in monetary units. H is the "optimum" selection index. By this we mean that for a given selection intensity, the response to selection will be greatest, in monetary units, if candidates for selection are ranked by H. Since the elements of y are generally unknown, the goal is to derive the linear index, I_s, of x, that is most similar to H. By "most similar" we mean either to maximize the correlation, or to minimize the mean squared deviation between I_s and H. Specifically, if b is defined as a vector of index coefficients, then $I_s = b'x$, and the objective is to solve for b that maximizes the correlation between b'x and a'y. Of course, like H, I_s will be a scalar in monetary units.

To derive I_s we will define three additional matrices, P, the n x n phenotypic variance matrix of the traits in x; C, the n x m genetic covariance matrix between the measured traits in x and the breeding values in y; and G, the m x m genetic variance matrix for the traits in y. The selection index coefficients are then derived from the following equation:

$$b = P^{-1}Ca \qquad [3.17]$$

Brascamp (1984) presents several methods to derive this equation. We will present only one method, based on minimizing the squared difference between I_s and H. This is also equivalent to maximizing the correlation between I_s and H, and maximizing the expected mean breeding value of individuals selected based on I_s. The derivation is simplified by assuming that both x and y are measured relative to their means. It is then necessary to minimize the following function:

$$(I_s - H)^2 = (b'x - a'y)^2 \qquad [3.18]$$

The expectation of the left-hand side of equation [3.18] can be computed as follows:

$$E(b'x - a'y)^2 = E(b'xx'b - 2b'xy'a + a'yy'a) \qquad [3.19]$$

since x and y are scored relative to their means, xx' and yy' will be the variance matrices for x and y, and xy' will be the covariance matrix between them. Thus:

$$E(b'x - a'y)^2 = b'Pb - 2b'Ca + a'Ga \qquad [3.20]$$

with all terms as defined above. Differentiating with respect to b and equating to zero we obtain:

$$2Pb - 2Ca = 0 \qquad [3.21]$$

$$Pb = Ca \qquad [3.22]$$

Solving for b we obtain equation [3.17]. If all traits included in the aggregate genotype are also included in the index, then G = C, and:

$$b = P^{-1}Ga \qquad [3.23]$$

This is the selection index equation most commonly presented.

3.5 Properties of the selection index

We have already noted that, of all possible linear indices of x, the selection index will have the highest correlation with H, and the lowest squared deviation. In addition, selection of individuals on I_s will result in maximum expected mean value for H of the selected individuals, and genetic response to selection on I_s will be greater than for selection on any other linear index of x. These and a few other properties of the selection index are summarized by Henderson (1973). We will now describe some additional useful properties of selection index, based on Cunningham (1969), James (1982), and Lin (1978).

From the above derivation, it should already be clear that the variance of the selection index can be computed as follows:

$$\sigma_{Is}^2 = \mathbf{b'Pb} = \mathbf{a'C'P^{-1}Ca} \qquad\qquad [3.24]$$

The variance of the aggregate breeding value will be $\mathbf{a'Ga}$. The covariance between I and H can be computed as follows:

$$\sigma_{H,Is} = \mathbf{a'yx'b} = \mathbf{a'Cb} = \mathbf{a'CP^{-1}Ca} = \sigma_{Is}^2 \qquad\qquad [3.25]$$

That is the variance of the index is also equal to the covariance between I_s and H. Since this is the case, the correlation between H and I_s, $r_{H,I}$ will be equal to $[\sigma_{Is}^2)/\sigma_H^2]^{0.5}$. This correlation for the selection index is parallel to the "accuracy" of single-trait genetic evaluation described in Chapter 1. Thus the response to selection on the index, ϕ_I, can be computed as follows:

$$\phi_I = i r_{H,I}\sigma_H = i\sigma_{Is} = i\sigma_{H,Is}/\sigma_{Is}^2 \qquad\qquad [3.26]$$

where i is the selection intensity, σ_H and σ_{Is} are the standard deviations of H and I_s, respectively, and $\sigma_{H,Is}$ is the covariance between H and I_s. ϕ_I will also be measured in monetary units. Thus the response to selection will be a direct function of the selection intensity and the standard deviation of the index.

As we will see in the following chapters, one of the major obstacles to the implementation of selection index is that the economic weights are not known, or change over region and time. We will deal now with the case of two alternative vectors of economic weights, $\mathbf{a_1}$ and $\mathbf{a_2}$. Using these two vectors, two different aggregate genotypes can be defined, H_1 and H_2. The similarity of these two breeding objectives can be measured by their genetic correlation, r_{H1H2}, which is computed as follows:

$$r_{H1,H2} = \frac{\mathbf{a_1'Ga_2}}{\sqrt{(\mathbf{a_1'Ga_1})(\mathbf{a_2'Ga_2})}} \qquad\qquad [3.27]$$

A question of major importance is the relative efficiency of selection on one index for improvement in another index. Because of the difficulty in the correct determination of economic values, it is important to estimate the consequence of selection on an index based on imprecise estimates of the true values. Using the previous notation, we will assume that a_1 is the true vector of economic weights, and a_2 is an alternative vector; and that I_1 is the optimum economic index, and I_2 is an alternative index, computed from the values in a_2. We can now ask what will be the efficiency of selection on I_2 to improve I_1? I_1 and I_2 can be considered two correlated traits. We showed above that the correlated response of one trait to selection on another can be computed from the regression of the breeding value of one trait on the breeding value of the other. In this case, the relative selection efficiency, RSE, will be the regression of the alternative index on the optimum index, which will be equal to the correlation between the two indices, which can be computed as follows:

$$\text{RSE} = \frac{b_1'Pb_2}{\sqrt{[(b_1'Pb_1)(b_2'Pb_2)]}} \qquad\qquad [3.28]$$

where b_1 and b_2 are the vectors of index coefficients derived from a_1, and a_2, respectively.

Optimum genetic response will be obtained when all traits with genetic correlations with the traits in the aggregate genotype are included in the index. If the ith trait of the aggregate genotype is deleted from the index, then the variance of the selection index will be reduced by b_i^2/w_i, where b_i^2 is the index coefficient for the ith trait, and w_i is the diagonal element for the ith trait in G^{-1} (Cunningham, 1969). From equation [3.26] we have that the response to selection is proportional to the standard deviation of the index. Thus the relative selection efficiency of the reduced index is computed as the ratio of the standard deviations of the reduced and complete indices, as follows:

$$\text{RSE} = \left[\frac{b_1'Pb_1 - b_i^2/w_i}{b_1'Pb_1} \right]^{1/2} \qquad\qquad [3.29]$$

If several traits are deleted from the selection index, say traits i to j, then the efficiency of selection of the reduced index can be computed by replacing b_i^2/w_i with $b_{i.j}'W_{i.j}^{-1}b_{i.j}$; where $b_{i.j}$ is the vector of index coefficients for the deleted traits, and $W_{i.j}$ is the appropriate submatrix of G^{-1} for the deleted traits.

Finally it is often of interest to compute the expected responses of the component traits to selection on the index. The genetic change for the ith trait due to selection on the index, ϕ_i, is computed as follows:

$$\phi_i = ib_{gil}\sigma_{Is} = i[Cov(g_i,I_s)/\sigma_{Is}^2]\sigma_{Is} = i[Cov(g_i,I_s)]/\sigma_{Is} \qquad [3.30]$$

where b_{gil} is the genetic regression of the ith trait on the index, $Cov(g_i,I_s)$ is the covariance between the genetic value of the ith trait and the index. $Cov(g_i,I_s) = Cov(g_i,p'b) = [Cov(g_i,p')]b$, where $Cov(g_i,p')$ is the ith column of C. Thus the vector of correlated responses for all traits, ϕ, is computed as follows:

$$\phi = iCb/\sigma_{Is} \qquad [3.31]$$

If all traits included in H are included in the index, then C can be replaced with G.

3.6 Example calculations

We will now return to the example of milk production and fat percent to illustrate an actual derivation of selection index. In this example we will assume that only these two traits have economic values and are to be included in the index. In this case the G and C matrices will be equivalent. The G and P matrices for this example are given above in Section 3.2. We will further assume that the economic value of a kg increase in milk product = \$0.3, and that the economic value of a one percent fat increase = \$1000. The index coefficients are then derived from equation [3.17] as follows:

$$P = \begin{bmatrix} 2{,}000{,}000 & -294 \\ -294 & 0.12 \end{bmatrix}^{-1} = \begin{bmatrix} 7.81\cdot10^{-7} & 1.9\cdot10^{-3} \\ 1.9\cdot10^{-3} & 13.02 \end{bmatrix} \qquad [3.32]$$

$$b = \begin{bmatrix} 7.81\cdot10^{-7} & 1.9\cdot10^{-3} \\ 1.9\cdot10^{-3} & 13.02 \end{bmatrix} \begin{bmatrix} 500{,}000 & -86.6 \\ -86.6 & 0.06 \end{bmatrix} \begin{bmatrix} 0.3 \\ 1000 \end{bmatrix} \qquad [3.33]$$

$$b = \begin{bmatrix} 0.1145 \\ 564 \end{bmatrix} \qquad [3.34]$$

Note that the ratio of the index coefficients is 1/4926, while the ratio of the economic weights is 1/3333. Fat percent is given greater weight in the selection index because its heritability is greater than milk production.

The variance of the index is derived from equation [3.24], and is equal to

26,422 dollars squared. Thus the standard deviation of the index is $162.5. The variance of the aggregate genotype, **a'Ga** is equal to 53,040 dollars squared. Its standard deviation is $230.3. The standard deviation of the index will always be less than or equal to that of the aggregate genotype. The correlation between the index and the aggregate genotype will be equal to the ratio of their standard deviations, which is equal to 0.706. Response to selection is computed as in equation [3.26], and will be equal to the selection intensity times $162.5. The responses of the individual traits are computed from equation [3.31], as follows:

$$\phi = \frac{i\mathbf{Gb}}{\sigma_{Is}} = \frac{i\mathbf{Gb}}{162.5} \qquad [3.35]$$

The elements of **Gb** are 8407.6 and 23.92. Thus the correlated responses per unit selection intensity are 51.73 kg milk and 0.1472% fat, respectively. Multiplying these values by the economic weights we obtain $15.5 for milk and $147.2 for fat percent, which sum to the value of $162.7 for the complete index. T h e relative efficiency of this index as compared to an alternative index can be computed from equation [3.28]. Assume an alternative index consisting of selection only on milk yield. In this case the economic value for fat percent will be zero, we will assume the same value of $0.3/kg milk. The correlation between this index and the optimum index, as derived from equation [3.27] will be 0.275. Thus the correlated response of the index to selection on milk alone will be equal to (0.275)($162.5) = $44.7 times the selection intensity.

Alternatively this result can be derived by computing the expected response to direct selection on milk, and the correlated response of fat percent. The response to selection on milk, as derived from equation [3.15] will be 353.5 kg milk, or $106 times the selection intensity. The correlated response of fat percent, as derived from equation [3.16] will be −0.0612 percent, or −$61.2. Thus the total response will be $106 − $61.2 = $44.8, which is nearly the same result derived above. (The discrepancy of $0.1 is due to rounding errors.) For comparison we note that the correlation between the optimum index and direct selection on fat percent is 0.603, and the expected economic response of selection on fat percent will be (0.603)($162.5) = $98. Therefore selection on fat percent is preferable to selection on kg milk, but both alternatives are much less effective than selection on the optimum index.

3.7 Summary

In this chapter we briefly reviewed the basic concepts of matrix algebra that are applicable to quantitative genetics. Matrix addition, subtraction, multiplication, and the inverse of a matrix were defined. Through matrix algebra it is possible to briefly denote and solve large systems of simultaneous equations. Phenotypic and genetic variance-covariance matrices were defined, and methods were developed to estimate the correlated response of a trait to direct selection on another trait.

The aggregate genotype is derived by multiplying the vector of each individual's breeding value for the traits under consideration by the vector of economic weights. The optimum linear selection index is defined as the linear index of trait values that maximizes the correlation between the index and the aggregate genotype. This index was derived for the situation in which the economic values are linear functions of the trait values. Basic properties of the selection index were derived, including expected response to selection. Equations were also derived for the genetic correlation between alternative indices, and the correlated response of the aggregate genotype to selection on an alternative index. These equations were illustrated through an example of selection for kg milk and fat percent in dairy cattle.

Chapter Four

Introduction to Systems Analysis

4.1 Introduction

In the first three chapters we discussed the relevant principles of quantitative genetics and economics, and showed how they can be combined with the aid of matrix algebra to obtain the optimum selection index. As stated previously, the main obstacle to application of selection index has been correct determination of the economic values of different traits. Although it may be relatively easy to compare the economic value of two products, such as wool and meat from sheep, it is not *a priori* clear how to determine the economic value of traits that affect the quantity of product, vs. traits that affect the cost of production. Production systems analysis provides a general framework for determination of breeding objectives. It requires that the objective be clearly specified, something that is generally not done in practice. Finally it provides a mechanism for decision making in relationship to the economics of the complete production scheme. In practice very little use has been made of systems analysis in animal breeding.

Successful use of systems analysis requires that three steps be completed: 1) the goals of the system have to be clearly defined, 2) the actual system has to be accurately represented by a model, and 3) changes in production programs must be implemented on the basis of the results of the analysis.

In this chapter we will discuss the role of systems analysis in animal breeding, describe the use of models, and the different methods used to derive solutions. This will be based mostly on Cartwright (1979), Dalton (1975), and Wilton (1979). Finally we will discuss some specific applications of systems analysis to animal breeding.

4.2 Defining goals for animal breeding

There are three main applications of systems analysis to animal breeding. In the previous chapter we defined the aggregate genotype as $H = y'a$. The first application will be to determine the values in a. That is to determine the objective of genetic selection, H, as opposed to the selection criteria, I. In the previous chapter, we defined a as the vector of economic values, and in the remainder of the book, we will generally assume that **a** is measured in monetary units per trait units. In the broadest sense, **a** is a vector of selection goals, which may be specified in monetary or energy terms or some other form of utility, as we discussed in Chapter 2. We showed how the optimum selection index, $I = x'b$, can be derived by maximizing the correlation between H and I. The vector **b** will be a vector of selection criteria, and will be a direct function of **a** and the genetic parameters. For example most breeding of dairy cattle has been directed to increasing 305-day mature equivalent milk production. As a selection criterion this trait is quite useful, since it is relatively easy to measure, and has significant heritability. However increasing this trait *per se* is clearly not the goal of selection, which could more appropriately be defined as maximizing lifetime profitability.

The second main application of system analysis will be the choice of breeding stock. This question can be considered either within a breed, or between breeds. The study of Sivaraysingan *et al.* (1984) is an example of the second type of decision. They used system analysis to optimize the semen selection of an individual dairy farmer under US conditions. System analysis can also be used to select the breed or breeds that best meet the goal of the system, and to decide whether these goals can best be achieved by selection within a breed or by crossbreeding. Finally, system analysis can be used to determine the optimum level of investment in breeding programs. To the best of our knowledge, no studies have as yet applied a systems analysis approach to this question.

4.3   The use of system analysis models and optimization techniques

Once the goals of the analysis have been defined, it is necessary to construct a model that accurately represents the system. Studies of systems can be carried out at various levels. Spedding (1975) lists four levels of understanding as: operation, repair, improvement, and construction. In animal breeding we will generally be interested with the last two levels, that is improvement of currently running systems, or construction of new ones. The desired properties of models are realism, precision, generality and resolution. In practice these properties conflict, and the system analyst must compromise, depending on the goals of the analysis. For example, as a model becomes more realistic and precise, it is likely to be less generally applicable. It is also possible to derive models with high generality and resolution, but either realism or precision must be sacrificed.

After the goals have been determined, and a model has been constructed, the appropriate systems analysis technique must be implemented. We will limit this discussion to optimization techniques. These techniques can be classified by three criteria: 1) linear vs. nonlinear, 2) static vs. dynamic, and 3) deterministic vs. stochastic. We will first discuss the simplest case of linear, static, deterministic programming, and then compare this to the other alternatives.

In linear programming (LP) the objective is to maximize an objective function subject to constraints. Both the constraints and the objective function are linear functions of possible "activities." Algebraically this problem can be phrased as follows:

Maximize: $z = c'x$ [4.1]

subject to: $Ax \leq k$,   [4.2]

$x \geq 0$ [4.3]

where z is the objective function, c is a vector of prices and costs, x is a vector of units of activities, k is a vector of level of resources, and A is a matrix of technical coefficients, such as estimated yields, feed requirements, etc. A, b, and c are fixed by the model, while x is variable. For example US dairy farmers can purchase semen from a large number of sires, each with a specific price and estimated transmitting ability. Sivaraysingan *et al.* (1984) used LP to economically rank semen of different sires. The objective function was profit, the units of activities were which cows would be inseminated by which sires, and the constraints were the level of resources, such as land, labor, and number of cows on a farm.

In the example given above, z is a linear function of x. If the objective function is a higher order function of the variable vector, then LP is no longer sufficient. We will use the example of Itoh and Yamada (1988) to illustrate an example of quadratic programming. Their problem was to find a vector of selection index weights, b, that would result in maximum genetic change for the index under the constraint of desired relative genetic gains not less than or not greater than certain levels for specific traits. Partitioning the traits into two groups, they required that the genetic change for the first group of traits be equal to k_1 and that the genetic change for the second group of traits be equal to or greater than k_2. These conditions are then met by the following set of equations:

$C_1'b = k_1$ [4.4]

$C_2'b \geq k_2$ [4.5]

$b'Pb$ minimum

where C_1 and C_2 are the covariance matrices between the phenotypic and genetic effects for the first and second group of traits, respectively, and P is the phenotypic variance matrix. Since the function to be minimized is a quadratic function of **b**, this is a quadratic programming problem. Theoretically the objective function could be some higher order function of the vector of variable activities, but in practice nearly all problems of static, deterministic optimization can be approximated with either a linear or quadratic function.

In the examples given above, all levels of activities were set at the same time, and the objective function is a direct function of these levels. Often we are interested with optimization of processes that change over time. Smith (1971) attempted to economically optimize the decisions involved in cow replacement in a dairy herd. Since decisions made in an earlier time will impact on the possible alternatives at a later time, the techniques of static programming described above are not applicable to this problem.

This problem was solved by a dynamic programming algorithm. In general terms, dynamic programming is a recursive methodology that begins at the final stage in the planning horizon and proceeds backward in time, stage by stage, until the present is reached. In the example given above, the final stage will be the composition of the herd at the end of the time period considered. The alternative decisions required to reach this state are then applied in reverse until the original herd composition is reached. If there is just one stage left in the process, returns will be realized during that stage, and possibly at the end of that stage in the form of salvage values. The goal is to optimize the possible activities so as to maximize the sum of these two sources of returns.

Generally dynamic programming problems will also be stochastic. That is, there is a possibility that different events will occur with differing probabilities. In the example of cow replacement, a cow at a given stage in her herdlife may either remain in the herd, or be culled for one of several reasons. Each of these events will have a given probability. This is of course different from the previous examples of deterministic programming where each vector of possible activities resulted in a single result for the objective function.

One important byproduct of system analysis modelling is identification of those parts of the system about which more information is required. That is, if small changes in certain elements included in the vector of activities result in major changes in the objective function, then it is likely that a key element of the model has not been sufficiently defined.

4.4 Definition of selection criteria

Very little has been written on the choice of selection criteria, as opposed to selection goals. This question is not considered within the context of selection index methodology, which assumes that the selection criteria, the vector of measured traits included in **x**, have been determined *a priori*. In fact the question of which traits to measure, and how these traits should be defined, is also a breeding decision. For example, in dairy cattle the main trait under selection has traditionally been total milk production from parturition through 305 days. Since production generally increases as the cow matures, production from younger cows is adjusted to expected mature production. Thus production past 305 days in milk is disregarded. Numerous studies have found that pregnancy has a negative effect on milk production. Thus several studies have suggested alternate selection criteria, such as annualized production (total lactation production divided by days between calvings) and days-open adjusted production. "Days open" is defined as the number of days between parturition and the following conception. Thompson, Freeman, and Berger (1982) compared days-open adjusted, annualized, and fat-corrected milk production records as alternatives to 305-day mature-equivalent first parity production. Although various comparisons were made between the different record types considered, the authors' choice as to the appropriate basis for comparison is not clearly stated.

A major dilemma in the determination of selection criteria is whether traits should be combined before or after genetic evaluation. To illustrate this problem, we will pursue the example of milk production. Assume that the objective of the selection is to maximize mean fat-corrected daily milk production over the cow's lifetime. Rather than select on first parity production, an alternative criterion, such as mean dairy production up to culling or date of evaluation, could be used. At the other extreme, we note that milk production is generally measured once monthly. Thus instead of combining the ten monthly scores into a single lactation record, each monthly test can be analyzed as a separate trait. The additional question of breeding for fat content in the milk was introduced in the previous chapter, in which we considered the optimal index for the two traits of milk production and fat percent. Rather than consider fat percent, we could have considered kg fat production. Breeding for the two traits of milk and fat production, it is still possible to increase fat concentration by giving milk production a negative economic value.

We will now try to elucidate the principles to determine the choice of selection criteria. First, optimally all traits measured with genetic correlations to the traits included in the selection objectives should be included in the selection criteria. In many commercial breeding programs traits that meet this condition are not included in the selection index. The usual justifications are that either these traits have low heritability or low economic value. Although it may turn out that once the optimal selection index is constructed, the index weights for certain traits may be negligible, it is difficult to predict this *a priori*. A number of studies have shown that including additional selection criteria with genetic correlations to traits in y will increase the rate of gain for the aggregate genotype (Hermas, Young, and Rust, 1987; Weller, 1989).

Second, it is generally advantageous to compute genetic evaluations on separate traits. An exception will be when the individual traits have nearly equal genetic parameters and economic value. Thus, combining the monthly milk production records into a single lactation record is probably justifiable, while combining lactation records is not. This will be explained by the latter example. Assume that first and second parity milk production have equal heritability, but unequal economic value; because first parity milk production occurs prior to second parity, and with a higher frequency. Thus by the theory of selection index, genetic gain is maximized by putting more emphasis on first parity. However, if the two traits are combined prior to evaluation, the relative emphasis becomes a function of the genetic and environmental parameters, rather than the economic weights.

The final principle we want to consider is the inadvisability of using ratios for selection criteria. Sokal and Rohlf (1969) present three disadvantages in the use of ratios. First, ratios tend to be less accurate than measurements on the component variables. Second, the distributions of ratios are often unusual, and may depart significantly from normality. Finally ratios do not provide information on possible relationships between the component traits. For example, assume that the goals of a beef breeding program are to increase unit calf weaned per unit weight of dam. Selection on this ratio might result in a genetic reduction in both variables. This result is probably not desirable, and can be avoided if selection is performed for a linear index of calf and dam weight, with a negative economic value for the latter trait.

4.5 Implementation of the results of the analysis

The decision maker must be recognized by the system analyst. If the decision maker cannot obtain results of the analysis readily, or does not accept the results of the analysis, implementation will not occur. Various entities are affected by animal breeding, and their goals will not be identical. This complicates definition of both breeding goals and the system to be analyzed. The major entities that must be considered are breeders, who may be either commercial or farmer cooperatives, farmers, food processors, merchants, consumers, and governments.

From this it is also clear that it is necessary to determine the level of the system being modeled, which in animal breeding can vary from a single animal to a national industry, and the time period to be considered. In an analysis on the national level the goal may be to increase the efficiency of production, while the goal of an analysis on the level of the individual farm may be to increase the farmer's profit. Even on the level of a regional breeding program, goals will be different if the breeding enterprise is commercial or cooperative. It should be emphasized that nearly all analyses of the effect of genetic improvement have been done either on the level of an individual animal, or on the level of an individual farm.

This anomaly was noted by Moav (1973), who discussed the question of who actually benefits from genetic improvement. He defined what he called the

"Progress-Surplus-Bankruptcy Cycle", which we first mentioned in Chapter 2. This cycle is illustrated in Figure 4.1. If higher productivity is confined to a small group of farmers, then higher productivity will not affect the supply curve, and profits of these farmers will increase. Most analyses of the effect of genetic improvement have assumed this to be the case. However in most cases, many producers will take advantage of genetic improvement. This was also illustrated in terms of supply and demand curves in Figure 2.11. Increased production will shift the supply curve to the left, creating disequilibrium between supply and demand. The demand curve will also shift to the left, due to rising affluence, but much less than the demand curve. Thus a new equilibrium point will be reached with a greater quantity being produced, but sold at a lower price. In a free market this is achieved via the "dynamic cobweb" described in Chapter 2, with the least efficient producers ceasing production. The remaining producers will then be able each to produce more, and presumably increase their profit somewhat. It would clearly be desirable to eliminate the left-hand side of the cycle of Figure 4.1. This may be done by government regulation. Unfortunately, in practice the effects of government intervention are often not as intended.

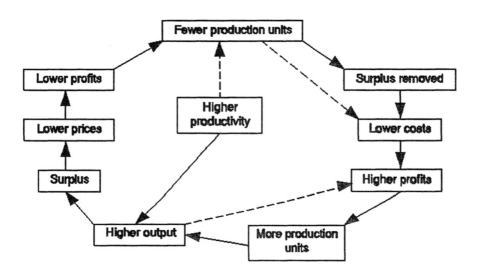

Figure 4.1. The "progressive-surplus-bankruptcy" cycle of Moav.

From the above discussion it is clear that the farmer generally does not benefit from genetic improvement. Thus defining breeding goals in terms of increasing the profit of the farmer may be unrealistic. It is then logical to ask who is the main beneficiary of genetic improvement? The answer is definitely not the breeder. Figure 4.2 illustrates this point. The line NE denotes the total gain from breeding to the national economy. If there is no competition among breeders then most of the gain from breeding may accrue to the breeder, and their mean profit will be a nearly linear function of the rate of genetic improvement. This situation is denoted by the B-C line. This of course is hardly ever the case, and competition will depreciate the value of breeding stock, so that the mean profit of breeders will tend to zero, as denoted by the B+C line. Thus semen from today's prize bulls can be purchased in a few years at a fraction of the current price. Generally consumers, rather than either the farmer or the commercial breeders are the main beneficiaries of genetic improvement.

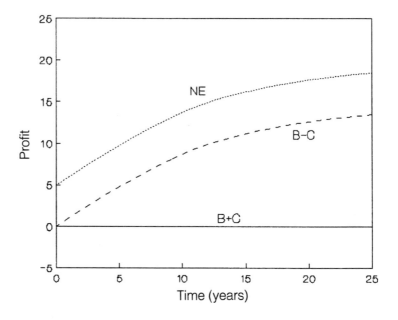

Figure 4.2. Depreciation of genetic gains due to competition. NE is the total gain from breeding to the national economy. B-C is the profit of breeders without competition. B + C is the profit of breeders with competition.

This clearly poses a problem for defining the goals of animal breeding, since the consumer does not make the major decisions that affect breeding programs. As a solution to this problem, analysis of breeding systems from the point of view of the national economy has been suggested. In this case it can be argued that the primary goal of animal breeding is to increase economic efficiency, defined as follows (Dickerson, 1970):

$$E = R/C \qquad\qquad [4.6]$$

where E is economic efficiency, R is returns from unit production, and C is the cost of unit production. This can be compared to profit, P, defined as follows:

$$P = R - C \qquad\qquad [4.7]$$

The difficulty with this approach is that even without genetic improvement, economic efficiency changes due to changes in the unit value of both income and expenses. Thus as an alternative, Maijala (1976) suggested biological efficiency. In this case the common denominator is biological energy. Biological efficiency is defined as the product, as measured in biological energy units, divided by cost of production in the same units. It is not clear though, how costs such as labor or rent could be factored into this calculation. Furthermore, from the point of view of system analysis, it does not appear that the relevant decision makers would desire to make decisions affecting breeding based on biological efficiency.

A further goal which should be considered is risk reduction. For example, even though maximization of E is desirable, a significant probability that income may fall below a certain level is intolerable. Thus a farmer may choose to use breeding stock with a lower estimated breeding value, but with a higher repeatability, if both are available at similar prices. In any case, if the purpose of the analysis is to aid the farmer or the breeder make optimum decisions from their point of view, clearly the goal must be to increase profit rather than efficiency. In later chapters we will try to resolve this problem somewhat.

The time period considered will also have a major impact on the results of the system analysis. As stated in Chapter 2, animal breeding is by its nature a long-term process, with most resources invested at the beginning of the program, while returns will keep accruing to infinity. It will also be shown in Chapter 6 and Part III that the effect of trait changes on both profit and economic efficiency will often be a function of the original trait values. Thus the optimum direction of selection will also be a function of the planning horizon.

4.6 Summary

Although in practice it has rarely been done, we believe that systems analysis should be used to determine breeding objectives. This should result in more realistic breeding goals rather than the commonly accepted goals of maximizing production, or profit for the individual farmer. Because of the different entities that interact with breeding decisions, neither of these goals is realistic in the long run for most practical breeding situations. The concepts of "vector of activities" and the "objective function" were explained, and the specific techniques used for optimization of the objective function were briefly described. Systems analysis first requires that the objectives be clearly defined. Second it provides a framework for consideration of breeding decisions within the total production framework. Third it puts breeding decisions into perspective with other management decisions.

PART II

ECONOMIC EVALUATION OF
GENETIC DIFFERENCES

In Part I we discussed the basic concepts necessary for the economic evaluation of animal breeding. In Part II we will discuss the methodologies that have been presented for the economic evaluation of genetic differences. In Chapter 5, we will discuss the main elements of costs and returns in economic evaluation of animal production. In the following chapter we will describe how genetic differences can be economically evaluated based on maximization of profit. We have already considered some of the deficiencies of profit as a criterion for genetic evaluation. In Chapter 7 we will discuss alternate methods of economic evaluation, including economic efficiency, biological efficiency, and return on investment, and will explain the conditions for equality of economic values by various criteria. Finally, in Chapter 8 we will consider long-term considerations of genetic improvement, a topic which must be considered for large animal

Chapter Five

The Main Elements of Returns and Costs

5.1 Introduction

Any economic evaluation should begin by considering two classes of variables, returns and costs. Often the concepts of returns and products have been confused in the economic evaluation of animal production. In a multi-enterprise production system, the returns of one enterprise may be quite different from the products that the consumer buys. For example in the production of poultry broilers, one enterprise might produce breeding stock, which is sold to a second enterprise in the form of chicks. These chicks are raised at the second enterprise which might sell the progeny as either eggs or chicks to a third enterprise that actually raises the broilers which are sold to the public. The same situation is common in beef production where one enterprise raises calves until weaning under range conditions, and a second fattens the calves under feedlot conditions.

Of course the costs will also be different for the different enterprises. Dickerson (1970) noted that the main costs of animal production for most species will be dependent on three main functions: 1) female production, 2) reproduction, and 3) growth of young. He excluded the costs related to male production, because for nearly all economically important species, this cost will be negligible, compared to the factors listed above. Thus economic efficiency, E, can be expressed by the following general equation for most production systems:

$$E = \frac{R}{C} = \frac{R_d + R_o}{EF_d + I_d + EF_o + I_o} \qquad [5.1]$$

where R_d is return from female production, R_o is return from offspring production, EF_d and I_d are feed and non-feed costs per dam, respectively; EF_o and I_o are feed and non-feed costs of her progeny; and the other terms are as defined previously. Calculations will generally be made on an annual basis for all terms. Various studies have preferred to estimate economic trait values based on the inverse of economic efficiency, which we will define as E_i. The reasons for this will be discussed in Chapter 7. If E_i is selected as the criterion for economic evaluation, then the goal will be to minimize E_i, as opposed to maximizing profit or E.

Generally speaking economic objectives will consist either of increasing returns or decreasing the costs of production. In this chapter we will consider the main elements of costs and returns that can be affected by breeding.

5.2 Elements of female production

The main animal products consumed are meat, milk, eggs and wool. It is possible that in the future, through genetic engineering, other products such as pharmaceuticals may become important, but in this discussion we will concentrate on the more traditional products listed above. Although the general equations presented for economic efficiency were derived by Dickerson (1970), we will use the notation of Moav (1973) with slight modifications, in the interest of consistency with the following chapters. In general, economic constants will be denoted with uppercase subscripted letters, and biological variables with subscripted x's, and other variables with other lowercase subscripted letters. The value of female production can be expressed with the following equation.

$$R_d = m_d x_D A_d \qquad\qquad [5.2]$$

where R_d is the yearly return per enterprise, due to female production, m_d is the number of females per enterprise, x_D is yearly volume of product/female and A_d is the value of product per unit volume. For example, assume a herd of 100 milk cows, each producing 8,000 kg milk/year, with a value of \$0.25/kg. R_d will be equal to $(100)(8,000)(0.25) = \$200,000$. Generally speaking, breeding has attempted to increase return by increasing x_D, although from the point of view of the producer, R_d could also be increased by increasing m_d or A_d. However, increasing m_d merely means increasing the size of the enterprise, and therefore is not relevant to breeding. A_d can be affected by changing the quality of the product. This is clearly important for most agricultural products, but in practice, much more emphasis has been put on increasing quantity, rather than quality of produce. There are two main reasons for this. First, measuring quality of a product is generally more difficult than measuring quantity. For example, quantity of milk produced can be scored by a simple scale, while measuring protein concentration requires at least a spectrophotometer. Second, there will generally be an antagonistic genetic correlation between quantity and quality of product. Continuing the previous example, both fat and protein concentration have negative genetic correlations with milk production.

Although most economic evaluations of breeding objectives have been done based on equation [5.2], it is inadequate for many situations. How does one compare milk production by goats and cows, or even compare milk production by different breeds which may differ markedly in size? To account for this factor it is sometimes useful to rewrite this equation as follows:

$$R_d = m_d x_{3d} x_{2d} A_d \qquad [5.3]$$

where x_{3d} is the mean weight of females and x_{2d} is production per unit female weight. It is now possible to consider whether the total enterprise is produced by a few big animals or many small ones. It should also be noted that total production, as computed in this equation can be increased by increasing either m_d, x_{2d} or x_{3d}. Various researchers have suggested that metabolic body size should be used rather than body weight. Metabolic body size is generally estimated as $x_3^{0.75}$. This value has been shown to be accurate over a large range of species.

Just as increasing the number of animals per enterprise is irrelevant to breeding, increasing the size of the production unit (animal) may not in fact increase either profit or economic efficiency. Unfortunately this point has been ignored by many studies that computed economic evaluations based on profit per animal.

5.3 Evaluation of female reproduction rate

Female reproduction rates differ markedly among domestic animals. This is illustrated by the examples in Table 5.1, from Moav (1973). Weight of dam, number of marketable offspring/yr, market weight per offspring, and reproduction ratio are listed for six species of domestic vertebrates. Reproduction ratio is defined as the ratio of total market weight of offspring per weight of dam. At one extreme are large mammals such as horses and cows with one progeny per year, and at the other extreme are fish and crustaceans with thousands of offspring per year.

Return from female reproduction can be evaluated by the following equation:

$$R_o = x_{1o} x_{2o} A_o \qquad [5.4]$$

where R_o is the return from offspring/year, x_{1o} is the number of offspring marketed/female/year, x_{2o} is the weight of offspring product, and A_o is the value per unit offspring product.

Table 5.1. Dam weight, number of marketable offspring/year, marker weight of offspring, and reproduction ratio for different domestic species.

Species	Weight of dam (kg)	No. of marketable offspring/year	Market weight per offspring	Reproduction ratio
Cattle	600	1	500	0.8
Sheep	60	2	40	1.3
Swine	200	15	100	7.5
Poultry	3	70	1.5	35
Turkeys	7	40	9	51.4
Carp (fish)	5	100,000	1	20,000

Breeding can increase R_o by increasing x_{1o}, x_{2o}, or A_o. Although we have designated A_o as an economic constant, there is generally some differential pricing based on the quality of product, which can be affected by breeding. Generally, the effect of breeding on R_o will be greatest by selecting for x_{2o}, rather than x_{1o} or A_o. The reasons for this are as follows. x_{2o} will generally be dependent on growth rate, which usually has high heritability and variance; while x_{1o} generally has low heritability, and A_o has low phenotypic variance, is generally difficult to measure, and as stated above, will be negatively correlated with x_{2o}. The number of offspring marketed is determined by several different genetically unrelated traits, such as interval between litters, number of offspring per litter, and juvenile mortality rates. Since these traits are related to natural fitness, they generally have low heritabilities. Furthermore selection for an index of several unrelated traits is inherently less efficient than selection for a single trait. In addition, as will be shown in Chapter 6, the economic importance of changes in x_{1o} decreases as the mean value of x_{1o} increases. Thus for animals with low reproduction rates, slight changes in x_{1o} will be of major economic importance, while for high fertility species, the economic importance of changes in this variable will be negligible. It should further be noted that for most domestic species, the coefficient of variation for x_{1o} increases with mean x_{1o}. Thus those species with the lowest reproductive rates, and therefore the highest economic value for this variable, have the lowest relative variance for this trait.

Despite these considerations, significant emphasis in selection has been devoted to increasing reproductive rate in most species. Moav and Hill (1966) give two reasons for this. In most cases one enterprise (which we will denote the breeder) produces juveniles or eggs, which are then sold to a second enterprise (which we will denote the rearer) that raises the animals for slaughter. The rearer will generally purchase young animals or eggs on a per unit basis. Thus the breeder will be primarily interested in the reproduction rate of his females. Although in theory the rearer should be willing to pay a premium price for a superior product, i.e. animals with a higher growth rate, in practice it is often difficult for him to evaluate the animals bought. Thus while a feedlot manager may be willing to pay a higher price for a Simmental calf than a Holstein, one day old chick looks just like any other.

The second reason has to do with the difference of estimation of profit for a constant vs. an expanding market, and will be explained in the next chapter.

5.4 Evaluation of feed costs

Feed costs can be divided into feed for the breeding female, and feed for the offspring. For each individual, feed costs can further be divided into feed for maintenance and for production. In the case of the breeding female, feed for production consists of the feed needed to produce offspring. Thus total feed costs of an integrated enterprise can be expressed by the following equation:

$$F_a = C_d m_d [x_{3d} F_{Md} + x_{1o}(F_{Pd} + x_{3o} F_{Mo} D + F_{Po} x_{2o})] \qquad [5.5]$$

Where F_a is the annual feed costs of the enterprise, C_d is the unit feed costs, x_{3d} is the metabolic body weight of the breeding female, F_{Md} is the maintenance feed required per unit metabolic body weight of the dam, F_{Pd} is the feed required by the dam per offspring produced, x_{3o} is the mean metabolic body weight of the offspring, F_{Mo} is the maintenance feed required/x_{3o}/day, D is the number of days from weaning to slaughter for the offspring, F_{Po} is the feed required per unit product, and the other terms are as defined previously. In this equation, x_{3o} is considered a biological variable, and D is considered a constant. This will be true for animals that are slaughtered at a constant age. However, if animals are slaughtered at a constant weight, then D will be a biological variable, and x_{3o} will be the economic constant.

Assuming slaughter at a constant age, the only terms in equation [5.5] that can be significantly affected by breeding are x_{3d}, x_{1o}, x_{3o} and x_{2o}, and increasing any of them will have a positive effect on F_a. Increasing m_d will have a proportional effect on both costs and returns. That is by changing m_d we merely change the size of the enterprise. The effect of changing x_{3d} will depend mainly on x_{1o} as illustrated above in Table 5.1. For large domestic animals, x_{1o} is relatively small, and changes in x_{3d} can have a significant effect on total feed costs. However for more prolific species feed for dams is negligible as compared to feed for progeny. The effect of changing x_{1o} will also depend on the mean of x_{1o}, as will be seen in the following chapter. Although from this equation, it would appear that breeding for reduction in mean offspring weight is a desirable goal, this is hardly ever done in practice. This is because there is generally a strong positive genetic correlation between x_{3o} and x_{2o}. This is of course evident when the main offspring product is meat, but will also be true for most other important products, such as milk, or wool. Finally feed costs can be reduced by decreasing x_{2_0}. However, since x_{2_0} is directly proportional to returns, unless profit is negative, it will be in the interest of the enterprise to increase x_{2_0}.

5.5 The relationship between growth rate and feed efficiency

For most domestic animals raised for slaughter, the main trait under selection is growth rate. This is because growth rate is usually highly correlated with feed efficiency. This will be illustrated by considering two cases, rearing to a constant slaughter weight, and rearing to a constant age. Assume that body weight increases linearly over time. This is approximately true for most domestic animals (Dickerson, 1970; Moav, 1973). Then x_{3_0} will be equal to the mean of initial and final body weight. Since differences in initial body weight are minimal, x_{3_0} will be equal to 1/2 final body weight, plus a constant. Rearing to a constant slaughter weight is illustrated in Figure 5.1. Body weight as a function of age is plotted for two growth weights. The integral of this curve will be equal to the product of x_{3_0} and D. In this case, increasing growth rate decreases D, but does not affect either x_{3_0} or x_{2_0}. Thus $x_{3_0}D$ is decreased, and feed efficiency is increased. This is the common situation for poultry production.

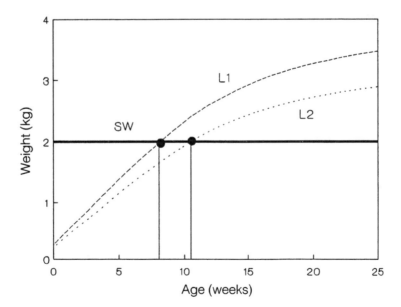

Figure 5.1. Effect of growth rate for slaughter at a constant weight. SW is slaughter weight. L1 and L2 are growth curves for two poultry strains.

Figure 5.2 illustrates the situation of slaughter at a constant age for two different growth rates. In this case, D is constant, but both x_{3_o} and x_{2_o} increase with increase in growth rate. Assuming that the initial weight is negligible compared to the final weight, we have the relationship that feed for maintenance is proportional to 1/2 final weight, while x_{2_o} is proportional to final body weight. The importance of this relationship can be illustrated as follows: If the number of offspring is doubled, with all other factors constant, then both maintenance feed and the quantity of meat produced will be doubled. However if growth rate is doubled, and all other factors remain constant, then x_{2_o} is still doubled, but maintenance feed increases only by 50%. Beef cattle are generally slaughtered at a constant age.

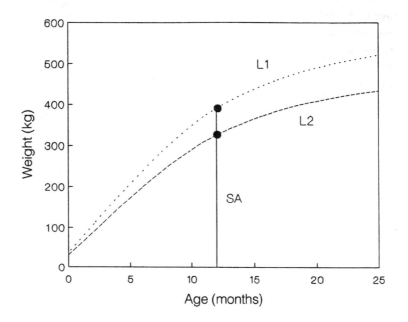

Figure 5.2. Effect of growth rate for slaughter at a constant age. SA is slaughter age. L1 and L2 are growth curves for two cattle strains.

From this discussion, it becomes apparent that, other factors being equal, there is an optimum slaughter age for all animals. Growth rates for most domestic animals are roughly linear until a given age, and then decline. Thus the optimum slaughter age can be found by expanding equation [5.1] as follows:

$$E = \frac{R_d + m_d x_{1o} x_{2o} A_o}{I_d + I_o + C_d m_d [x_{3d} F_{Md} + x_{1o}(F_{Pd} + x_{3o} F_{Mo} D + F_{Po} x_{2o})]} \qquad [5.6]$$

We will now define the following approximate equalities:

$$x_{3o} = x_{2o}/2 \qquad [5.7]$$

$$x_{4o} = x_{2o}/D \qquad [5.8]$$

$$K_1 = (I_o + I_d)/m_d + C_d(x_{3d} F_{Md}) \qquad [5.9]$$

$$K_2 = C_d F_{Pd} \qquad [5.10]$$

$$K_3 = C_d F_{Mo} \qquad [5.11]$$

$$K_4 = C_d F_{Po} \qquad [5.12]$$

where x_{4o} is growth rate, K_1 is costs independent of progeny weight, K_2 is additional dam feed costs per progeny, K_3 is maintenance costs per kg progeny, K_4 is cost of production of kg progeny, and the other terms are as defined previously. Substituting these terms into equation [5.6], we obtain:

$$E = \frac{x_{3d} x_{2d} A_d + x_{1o} A_o x_{4o} D}{K_1 + K_2 x_{1o} + (K_3 x_{1o} x_{4o} D^2)/2 + K_4 x_{1o} x_{4o} D} \qquad [5.13]$$

Assuming that return per dam is negligible as compared to return from progeny and that additional dam feed costs are negligible as compared to other feed costs, simplifying and inverting we obtain the following equation:

$$1/E = \frac{K_1/x_{1o} + (K_3 x_{4o} D^2)/2 + K_4 x_{4o} D}{A_o x_{4o} D} \qquad [5.14]$$

Differentiating with respect to D and equating to zero gives:

$$D_{max} = [2K_1/(x_{1o} K_3 x_{4o})]^{0.5} \qquad [5.15]$$

where D_{max} is optimum slaughter age. Thus D_{max} will increase as a function of costs independent of progeny weight, and decrease as a function of number of progeny per dam, growth rate, and maintenance costs per kg progeny.

Equation [5.15] will be applicable only if D_{max} falls within the linear growth phase. This will be the case for poultry broiler production, but not for beef production. As animals approach maturity, growth rates decline, and fat production, which requires more energy for production than muscle, also increases. Thus the optimum slaughter time for beef calves is near the onset of sexual maturity. The economical efficiency of beef production can therefore be increased by extending the linear growth phase. This is the main difference between large and small beef cattle breeds.

There is probably economically important genetic variance for feed efficiency between individual animals after correction for differential growth rates. Although individual feed consumption has a high genetic correlation with growth rate, it is still less than unity. This relationship has been shown both for poultry and beef cattle (Pym and Nicholls, 1979; Weller, Quaas, and Brinks, 1990). Furthermore both traits have high heritability. However, selection for increased feed efficiency independent of growth rate requires that individual feed intake be measured, and this is prohibitively expensive under commercial growth conditions for all domestic species.

5.6 Non-feed costs of production

Although the major production costs will be feed-related, there will also be significant non-feed costs. These can be divided into fixed costs per enterprise, per breeding female, and per progeny. The major non-feed fixed costs will be labor, rent, interest, buildings, veterinary costs, and replacement breeding females. The only element of these costs that can be directly affected by breeding is veterinary costs. Even though disease-related costs are significant, relatively little emphasis has been devoted to breeding animals for disease resistance, because of poor recording and generally low heritabilities. Although the other elements of non-feed costs can generally not be affected by breeding, we will see in the following chapters that they will affect the calculation of the economic evaluation of genetic differences. As is the case for feed costs, non-feed costs that are a function of the number of breeding females will be relatively more important for low fertility species.

5.7 Summary

In this chapter we discussed the main elements of costs and returns in animal production. Efficiency of production or profit can be increased either by decreasing costs or increasing returns. The main elements that can be affected by breeding are the number of offspring per breeding female, the quantity of product per offspring, the quality of the product, the body weight of the breeding female, and the growth rate of her offspring. Except for the body weight of the dam, increasing the other factors will affect both costs and returns. Most of the emphasis in selection has gone into increasing quantity of product per offspring, including growth rate and number of progeny per breeding female.

Chapter Six

Evaluation of Genetic Differences from Profit Equations

6.1 Introduction

In Chapter 3 we defined the economically optimum selection index as the index that maximizes the economic gain from selection. We showed how the principles of selection index can be used to derive the economically optimum index, provided that the genetic parameters and the economic values of the different traits are known. In the previous chapter we discussed the main elements of costs and returns of animal products, within the context of economic efficiency, but did not elaborate on the actual criteria for determining the economic value of a unit change in each trait.

In Chapter 4 we discussed the three main criteria that have been suggested to evaluate genetic differences: profit, economic efficiency, and biological efficiency, and the advantages and disadvantages of each. In this chapter we will explain how to evaluate differences between individuals or strains for traits of economic importance based on profit. In general terms this is accomplished by expressing profit as a function of the component traits. The economic values of the traits are then computed as the partial differentials of these traits with respect to profit. The notation and most of the examples will be based on Moav (1973). We will show that the estimation of marginal profit can be quite complex under certain circumstances, and will depend both on the characteristics of the traits under selection and the market constraints. Alternative criteria for evaluating genetic differences will be discussed in Chapter 7.

6.2 The basis for evaluation of trait differences

In order to construct profit equations, it is necessary to first consider the unit of comparison. For example, we can consider profit per unit product, per production unit (animal), per unit animal weight, per enterprise (farm), or for the entire national economy. At first glance, this question may not seem important. The reader may consider this analogous to asking whether a trait is measured in grams or pounds. In fact it will be demonstrated that radically different results can be obtained, depending on the unit selected as the basis of evaluation.

We will start with the example of egg production in poultry. Assume that the objective is to compute the economic value of a unit change in the number of eggs laid per hen. At the beginning we will assume that all costs and returns of the layer mother are negligible compared to the costs and returns of the layer. Profit per unit of product, in this case profit per egg produced, is computed as follows:

$$P_1 = A_1 - F_1 - V(x_1) = K - V(x_1) \qquad [6.1]$$

Where P_1 is profit/egg, A_1 is income/egg, F_1 is fixed costs/egg, x_1 is the number of eggs/hen, and $V(x_1)$ is the variable costs of egg production. F_1 and $V(x_1)$ will include both feed and non-feed costs. $V(x_1)$ denotes that the variable costs of egg production are some function of x_1. Since both A_1 and F_1 are independent of x_1, they can be combined into a single constant denoted K in the right-hand-term of equation [6.1]. (Note that this definition of fixed and variable costs is different from the definition in Chapter 2. In that chapter fixed costs were defined as costs independent of the amount produced, while variable costs were defined as all other costs.)

In the previous chapter we explained that it is convenient to divide costs into feed and other costs. In equation [6.1] feed costs included in F_1 will be the feed required to produce eggs, while other feed costs will be included in $V(x_1)$. Similarly non-feed costs that are a direct function of the number of eggs produced, such as egg handling labor, will be included in F_1; while other non-feed costs will be included in $V(x_1)$.

In order to obtain a simple algebraic expression for $V(x_1)$, we will assume that all costs not included in F_1 are a direct function of the number of layers. Then equation [6.1] can be rewritten as follows:

$$P_1 = A_1 - F_1 - F_2/x_1 = K - F_2/x_1 \qquad [6.2]$$

Where F_2 is the fixed costs per hen, and the other terms are as defined above. In this equation profit is now expressed as an inverse function x_1. Increasing x_1 increases profit/egg by distributing the fixed costs per hen over a greater number of eggs.

The marginal change in profit/egg/hen (the a-value of the selection index) is computed by differentiating equation [6.2] with respect to x_1 as follows:

$$\frac{d(P_1)}{d(x_1)} = \frac{F_2}{(x_1)^2} \qquad [6.3]$$

Equation [6.3] is possibly the most important equation in this book so far. We note first that as long as x_1 is positive, the change in profit, per added egg/hen will be positive. However, the marginal increase in profit is *not* a constant, but rather a nonlinear function of x_1. In fact, as the number of eggs/hen increases, that additional profit/egg decreases. This equation points out one of the main difficulties in application of selection index; namely, that the economical values of the traits under selection are often functions of the phenotypic trait values.

We will now rewrite equation [6.2] to evaluate profit per hen. This can be done by multiplying both sides of equation [6.2] by x_1.

$$P_2 = x_1(P_1) = K(x_1) - F_2 \qquad\qquad [6.4]$$

Where P_2 is profit per hen, and the other terms are as described above. Differentiating this equation with respect to x_1 yields:

$$\frac{d(P_1)}{d(x_1)} = K \qquad\qquad [6.5]$$

That is profit/hen is a linear function of x_1, and the marginal change in profit (the a-value of the selection index) is now a constant. Thus the economic value of a unit change in the number of eggs per hen will be different if profit is computed per hen, or per egg. Furthermore, in the former case, the economic value will be a constant, while in the later case it will be an inverse function of x_1. We will consider other bases for profit calculation in the later sections of this chapter.

6.3 Multiple-trait economic evaluation

We will now consider the case of simultaneous economic evaluation of several traits. In the example given above, in addition to number of eggs/hen, mean weight of eggs and hen body weight will be important economic traits. Following the notation of Moav (1973) these two additional traits will be denoted x_2 and x_3, respectively. We will first assume that x_2 is constant and compute the economic value of the two remaining traits on profit. The fixed costs per hen can now be computed as follows:

$$F_2 = (K_4 + K_3x_3) \qquad\qquad [6.6]$$

where K_4 is the fixed cost per hen, K_3 is the fixed cost per unit weight of hen, and the other terms are as defined previously. Substituting equation [6.6] into equation [6.4], profit per hen can now be expressed by the following equation:

$$P_2 = x_1[K - (K_4 + K_3x_3)/x_1] \qquad\qquad [6.7]$$

with all terms as defined above. Substituting equation [6.6] into equation [6.2], profit per egg can be computed as follows:

$$P_1 = P_2/x_1 = K - (K_4 + K_3x_3)/x_1 \qquad [6.8]$$

Finally we can also compute profit per gram hen, P_3, by dividing equation [6.7] by x_3 as follows:

$$P_3 = P_2/x_3 = (x_1K)/x_3 - K_4/x_3 + K_3 \qquad [6.9]$$

with all terms as defined above. The economic values of x_1 and x_3 are the partial differentials of these variables with respect to profit. These values are summarized in Table 6.1 for the three profit criteria in equations [6.7] through [6.9].

Table 6.1. Partial differentials of profit with respect to eggs/hen (x_1) and hen body weight (x_3).

Profit criteria	Partial derivatives	
	x_1	x_3
Per egg (P_1)	$[K_4 + K_3x_3]/x_1^2$	$-K_3/x_1$
Per hen (P_2)	K	$-K_3$
Per gram hen (P_3)	K/x_3	$[K_4 - Kx_1]/x_3^2$

We first note that the economic values for both traits will be different for each of the three profit criteria. Under the assumptions that K, K_3 and K_4 are all positive, and that x_1 and x_3 are greater than unity, then the economic value of x_1 will be greater if profit is computed per hen than if profit is computed per gram hen. Likewise the absolute economic value of x_3 will be greater if profit is computed per hen, as opposed to per egg. We further note that the economic values are equal to constants only in the case of profit per hen. Thus linear selection index cannot be directly applied for either of the other two profit criteria.

6.4 Choice of the appropriate profit criteria

Most studies that have attempted to evaluate genetic differences, have done so by the criterion of profit per animal. This criterion is probably justifiable only under a very short-term profit horizon. For example it may be difficult for a dairy farmer to significantly change the number of cows in his herd within a week, but there is no reason that he cannot appreciably change this number over a space of several months or years. Two alternative constraints which will apply both in the short- and long-term are constraints on production or constraints on investment. We will first consider the case of constraints on production.

In order that production should not exceed demand, most developed countries have imposed production quotas on many agricultural products. If each enterprise has a production quota, then production will be a fixed quantity for both the enterprise and the national economy. We will now compute profit per enterprise, P_E, as profit per animal, times m, the number of animals raised:

$$P_E = mP_2 = mx_1P_1 = Q(P_1) \qquad [6.10]$$

where Q, the quantity of the demand, is equal to m times x_1, and the other terms are as defined above. At equilibrium, then Q will also equal the quantity of the supply, as shown in Chapter 2. (We have designated the product of m and x_1 as "demand" rather than supply because it is demand that we assume to be fixed.)

With fixed Q, an increase in x_1 will cause a reduction in m. Thus m can be computed as a function of x_1 and Q as follows:

$$m = Q/x_1 = (m_0x_{10})/x_1 \qquad [6.11]$$

where m_0 and x_{10} are the original values for m and x_1 prior to the change in x_1. Profit for fixed demand, P_Q can then be expressed as follows:

$$P_Q = m_0x_{10}P_1 = m_0X_{10}[K - (K_4 + K_3x_3)/x_1] \qquad [6.12]$$

Note that the only variables in this equation are x_1 and x_3. Therefore, since m_0x_{10} is a constant, the profit equation in [6.12] is proportional to profit per egg in equation [6.8]. Thus the partial derivatives of this equation will be equal to the partial derivatives of equation [6.8], multiplied by the constant, m_0x_{10}.

We will now consider the other two possibilities of profit for a fixed number of animals (production units) and profit for a fixed total weight of animals. The latter alternative can be considered approximately equal to profit for fixed investment. Profit for a fixed number of animals, P_M, is computed as follows:

$$P_M = m_0P_2 = m_0x_1[K - (K_4 + K_3x_3)/x_1] \qquad [6.13]$$

with all terms as defined previously. This is of course profit per hen, multiplied by the constant, m_0.

For the case of fixed investment, we will require that the total weight of hens should be fixed. That is:

$$W = mx_3 = m_0x_{30} \qquad [6.14]$$

where W is total weight of hens (investment), x_{30} is the initial hen weight, and the other terms are as defined previously. From this equation we see that with fixed weight of hens, x_3 will be an inverse function of m. Profit with fixed investment, P_W, is computed as follows:

$$P_W = mx_3P_3 = m_0x_{30}x_1P_1/x_3 = m_0x_{30}[(x_1K)/x_3 - K_4/x_3 - K_3] \qquad [6.15]$$

With all terms as previously defined. As in the previous cases, P_W is equal to P_3 times m_0x_{30}, which is a constant.

Since the objective of breeding is to increase profit, we need consider chiefly those situations that result in increased profit relative to the original situation, specifically $x_1 > x_{10}$, and $x_3 < x_{30}$. Within this parameter space we can then deduce the following inequality:

$$P_W > P_M > P_Q \qquad [6.16]$$

This relationship can be explained as follows: For P_Q profit can be increased only by decreasing costs per unit product, for P_M, profit can also be increased by increased production, and for P_W it is possible to further increase profit by decreasing the production unit with fixed investment. The partial differentials for these three profit criteria are listed in Table 6.2. As should be clear from the previous discussion, the values for each row in Table 6.2 are proportional to the corresponding row in Table 6.1.

In Chapter 3 we showed that multiplication of the vector of economic values by a scalar will have the same effect on the index weights. This is equivalent to changing the scale of measurement for the economic values. For example, if the economic values and index coefficients are measured in dollars/kg, multiplication of the index coefficients by 2.2 changes the scale to dollars/lb, but does not change the ratios among the economic values. Thus the *ratios* among the economic weights are more important than their actual values. We have therefore also included the ratio of the partial derivatives in this table. These ratios are also different for the three profit criteria, and except for P_M they are also functions of the trait values. In conclusion we see that the profit criteria can have a marked effect on the economic values of the traits included in a selection index.

Table 6.2. Partial differentials of profit with respect to eggs/hen (x_1) and hen body weight (x_3) by three different enterprise criteria.

Profit criteria	Partial derivatives		
	x_1	x_3	ratio $x_1:x_3$
Fixed demand (P_Q)	$\dfrac{m_0 x_{10}[K_4 + K_3 x_3]}{x_1^2}$	$\dfrac{-m_0 x_{10} K_3}{x_1}$	$\dfrac{-[K_4 + K_3 x_3]}{x_1 k_3}$
Fixed number of production units (P_M)	$m_0 K$	$-m_0 K_3$	$-K/K_3$
Fixed investment (P_W)	$\dfrac{m_0 x_{30} K}{x_3}$	$\dfrac{m_0 x_{30}[K_4 - K x_1]}{x_3^2}$	$\dfrac{-K x_3}{K x_1 - K_4}$

6.5 Differential production quotas

We will now consider an example of profit computed as a function of three traits. In addition to the two previous traits of eggs/hen and hen weight, we will add a third trait of egg weight, x_2. If eggs are priced by weight, then this variable will affect both income and costs. Profit per hen can now be expressed as follows:

$$P_2 = K_1 x_1 x_2 - K_2 x_1 - K_3 x_3 - K_4 \qquad [6.17]$$

where K_1 is income per gram egg less fixed costs per gram egg, K_2 is fixed costs/egg, and the other terms are as defined previously. As in the previous discussion, profit per egg can be computed by dividing equation [6.17] by x_1, while profit per gram hen can be computed by dividing this equation by x_3. In addition it is now possible to define a fourth profit criterion, namely profit per gram egg, which can be computed by dividing equation [6.17] by $x_1 x_2$. The coefficients of the four constants $K_1 - K_4$ are summarized in Table 6.3.

As in the previous example, the economic values of the three traits can be computed as the partial derivatives of each profit criterion with respect to each trait. These values are given in Table 6.4. As in the two-trait case, the partial differentials are quite different, depending on the profit criteria. Since the partial differentials for profit per egg and profit per gram egg are also different, which criteria are appropriate for conditions of fixed demand? The answer will depend on how fixed demand is determined. For example, if each producer has a production quota computed in number of eggs, but is paid by egg weight, then the proper criterion would be profit per egg. Note that in this case the economic value of egg weight is K_1. That is with respect to weight of eggs, the producer is effectively in an unconstrained market, and it will be to his advantage to put most of the emphasis of selection on increasing egg weight. However, if each farmer receives a quota in weight of eggs produced, or one considers the viewpoint of the national economy, then the proper criterion will be profit per weight of eggs produced.

Table 6.3. Coefficients of the economic constants with four different profit criteria.

Profit criteria	Coefficients			
	K_1	K_2	K_3	K_4
Per egg (P_1)	x_2	1	x_3/x_1	$1/x_1$
Per hen (P_2)	$x_1 x_2$	x_1	x_3	1
Per gram hen (P_3)	$x_1 x_2 / x_3$	x_1 / x_3	1	$1/x_3$
Per gram egg (P_4)	1	$1/x_2$	$x_3/(x_1 x_2)$	$1/(x_1 x_2)$

Another example of this problem is calculating the economic weights for components of milk production. The economically important components of whole milk are butterfat, protein, and lactose. Both total milk produced and component concentration can be affected by both breeding and management. Currently most developed countries pay a price differential based on protein and fat concentration. In addition, the energy requirements to produce these components are not equal. It requires more energy, and therefore costs more, to produce a gram of fat than a gram of protein, and production of a gram of protein costs more than an equal weight of lactose. If production quotas are in kg fluid milk, while a price differential is paid for protein and fat production, then the added profit for additional production of these components may be much greater than from additional milk production.

Table 6.4. Partial differentials of profit with respect to eggs/hen (x_1), egg weight (x_2) and hen body weight (x_3).

Profit criteria	Partial derivatives		
	x_1	x_2	x_3
Per egg (P_1)	$\dfrac{K_3x_3 + K_4}{x_1{}^2}$	K_1	$\dfrac{-K_3}{x_1}$
Per hen (P_2)	$K_1x_2 - K_2$	K_1x_1	$-K_3$
Per gram hen (P_3)	$\dfrac{K_1x_2 - K_2}{x_3}$	$\dfrac{K_1x_1}{x_3}$	$\dfrac{K_2x_1 + K_4 - K_1x_1x_2}{x_3{}^2}$
Per gram egg (P_4)	$\dfrac{K_3x_3 + K_4}{x_1{}^2x_2}$	$\dfrac{K_2}{x_2{}^2} + \dfrac{K_3x_3 + K_4}{x_1x_2{}^2}$	$\dfrac{-K_3}{x_1x_2}$

6.6   Graphical representation of profit: reproductivity vs. productivity

The relationships described above can also be represented graphically by plotting one trait as a function of a second trait for a given profit level. If this function is plotted for a number of different profit levels, then the figure is denoted a "profit map", and the curves for the individual profit levels are denoted "profit contours". For example, x_3 in equations [6.12] or [6.15], can be plotted as a function of egg number and profit for fixed number of eggs (demand) or fixed weight of hens (investment). Solving for x_3 from these equations we obtain:

$$x_3 = \frac{1}{K_3} \left[\frac{x_1(m_0x_{10}K - P_Q)}{m_0x_{10}} - K_4 \right] \qquad [6.18]$$

$$x_3 = \frac{m_0x_{30}(Kx_1 - K_4)}{P_W + m_0x_{30}K_3} \qquad [6.19]$$

Since the other terms are constants, x_3 is now expressed as a function of x_1 and profit.

The profit maps derived from equations [6.18] and [6.19] are plotted in

Figure 6.1 for the constant values of Moav (1973). The solid lines represent the profit contours for fixed demand (fixed number of eggs), and the broken lines represent the profit contours for a fixed number of hens. Since profit is an inverse function of body weight, the scale of body weight is inverted. Thus on this graph profit is maximum at the upper right-hand corner, and minimum at the lower left-hand corner. This convention will be followed throughout. By both criteria, x_3 is a linear function of x_1. Thus the profit contours are straight lines by both profit criteria. However the profit contours are not parallel. Furthermore, only the zero profit contour is congruent by both criteria. Thus if individuals are ranked for selection based on their expected profit, the ranking will be different for different criteria.

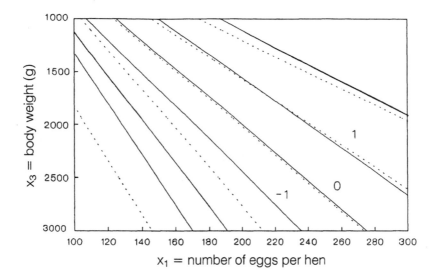

Figure 6.1. Profit map for laying hens. Body weight is a function of number of eggs per hen for the constant values of Moav (1973). The solid lines are the profit contours for fixed demand (fixed number of eggs), and the broken lines are the profit contours for a fixed number of hens. Since profit is an inverse function of body weight, the scale of body weight is inverted. Profit contour units are IL 10^4 per enterprise.

Other things being equal, profit will be maximized by moving at right angles to the current profit contour. Since the profit contours for a given criterion are not parallel, the direction of maximum profit will change as profit increases. Furthermore, since the profit contours computed by the two criteria are also not parallel, except at zero profit, the direction of change for maximum profit at a given profit level will also depend on the profit criterion.

In the previous chapter we explained that for most species, cost can be partitioned into costs of production, and costs of female reproduction. Until now we have only considered the first element. In addition to the cost involved in keeping the laying hens, there will also be costs of keeping the mother hens that produce the laying hens. For an integrated enterprise that raises both mother hens and layers, profit can be expressed by the following equation:

$$P = K - V_2 - V_1 \qquad\qquad [6.20]$$

where V_2 represents the variable costs of production, V_1 represents the variable costs of reproduction, and K is return per unit production less fixed costs per unit production. We will now expand this equation following Moav (1973), for the specific example of pig production in an integrated enterprise that raises both sows and pigs for slaughter.

$$P_1 = K_1 - K_2 x_2 - K_3/x_1 \qquad\qquad [6.21]$$

where P_1 is profit per pig marketed, x_1 is number of pigs weaned per sow per year, x_2 is age to a fixed market weight, K_1 is income less costs independent of x_1 and x_2, K_2 is costs dependant on x_2, and K_3 are fixed costs (feed and non-feed) per sow. x_2 can also be defined as the food conversion ratio growth rate. In several previous equations, profit was also an inverse function of x_1. Note, however, the difference between this equation and equation [6.8]. The importance of this difference will become apparent shortly.

In the previous chapter we explained how increasing growth rate will also increase feed efficiency. In equation [6.21] we assume that pigs are marketed at a constant weight. Thus increasing growth rate reduces expenses by decreasing the number of days that the pig must be fed prior to slaughter. For simplicity this function is assumed to be linear. Similar to the previous examples, we will now consider profit per fixed demand (pigs marketed), P_D; and fixed number of production units (sows), P_M. These equations are derived in a parallel manner to equations [6.12] and [6.13]:

$$P_Q = m_0 x_{10} P_1 = m_0 x_{10}(K_1 - K_2 x_2 - K_3/x_1) \qquad\qquad [6.22]$$

$$P_M = m_0 x_1 P_1 = m_0(K_1 x_1 - K_2 x_1 x_2 - K_3) \qquad\qquad [6.23]$$

where m_0 is the number of sows/enterprise, x_{10} is the original number of pigs/sow, and the other terms are as defined previously. The profit contours can then be computed by solving for x_2 as a function of profit and x_1, as follows:

$$x_2 = \frac{1}{K_2} \left[K_1 - \frac{K_3}{x_1} - \frac{P_Q}{m_0 x_{10}} \right] \qquad [6.24]$$

$$x_2 = \frac{1}{K_2} \left[K_1 - \frac{K_3}{x_1} - \frac{P_M}{m_0 x_1} \right] \qquad [6.25]$$

with all terms as defined previously. The profit contours for these functions are given in Figure 6.2 for the constant values of Moav (1973) for a swine enterprise.

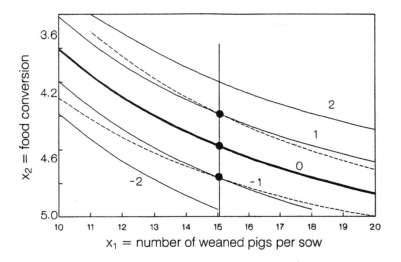

Figure 6.2. Profit map for a swine enterprise for the constant values of Moav (1973). x_2 is plotted on a reverse scale, because of the negative relationship between x_2 and profit. Profit contours for fixed demand and fixed number of sows are denoted with solid and broken lines, respectively. Profit contour units are IL 10^3 per enterprise. Vertical line is the initial value for x_1.

As in Figure 6.1, x_2 is plotted on a reverse scale, because of the negative relationship between x_2 and profit. Profit contours for fixed demand and fixed number of sows are denoted with solid and broken lines, respectively. Note first that in both equations, x_2 is an inverse function of x_1. Therefore the profit contours are nonlinear functions. As in Figure 6.1, the profit contours are congruent only when $P_M = P_Q = 0$. The profit contours with $P_M = P_Q$ cross at $x_1 = x_{10}$. That is, if the number of pigs per sow remains constant, then profit by both criteria will be equal for any value of x_2. As in the previous example, animals will be ranked differently by these two profit criteria.

The significance of the nonlinearity will be two-fold. First we will consider the effect of changes in x_2 as a function of x_1. At any combination of values for x_1 and x_2, the effect on profit of a unit change in x_2 will be equal. However, for P_Q and a constant value for x_2, a unit change in x_1 will have a greater effect on profit at a low number of pigs than at high number. This relationship is of course evident from the partial derivatives of equation [6.22]. Second, as stated above, profit is increased most rapidly by progressing at right-angles to the profit contours. In the example in Figure 6.1 the direction of maximum increase in profit will be parallel for all points along a profit contour. In Figure 6.2, for points with equal profit, the direction of maximum increase in profit will be different. We will return to these points when we consider nonlinear selection indices in Chapter 9, and evaluation of crossbreeds in Part V.

6.7 Summary

In this chapter we showed how the economic values of traits can be computed as the partial derivatives of profit equations. These equations point out two major difficulties in the practical application of selection index. First, the economic values are often functions of the current trait values; and second, the economic values will depend on the criteria used to compute profit. In general terms, the profit criteria considered were profit per fixed number of production units (animals), fixed demand, and fixed investment (weight of animals). We showed how profit contours can be used to display these relationships graphically for two traits. Profit contours will be linear if two production traits are compared, but nonlinear if a productive and reproductive trait are compared. Consequences of these relationships will be explored in Chapter 9 and Part V.

Chapter Seven

Evaluation of Genetic Differences by Alternate Methods

7.1 Introduction

In Chapter 3 we explained the principles of selection index, and showed how genetic progress can be maximized economically if the economic values of the traits under selection are known. In the previous chapters we considered the main elements of costs and returns, and showed how unit changes in specific traits of economic importance can be evaluated economically based on maximization of profit. In Chapter 4, we discussed some of the disadvantages of using maximization of profit as the criterion for economic evaluation, and other disadvantages became apparent in the previous chapter. Therefore other criteria for economic evaluation have been suggested in the literature. The main alternatives to profit are economic efficiency, biological efficiency, and return on investment.

In this chapter we will discuss the advantages and disadvantages of these methods, as compared to profit, and will explain in detail the conditions for equality between different profit criteria and economic efficiency. It will be seen that these conditions are general enough that the problem of differing economic values for different profit criteria is less serious than thought originally. Finally we will consider empirical methods for estimating economic values, based on actual prices and field data.

7.2 Economical and biological efficiency and return on investment as alternative criteria to profit for economic evaluation of trait unit changes

In Chapter 4 we defined profit (net income) and economic efficiency as follows:

$$P = R - C \qquad\qquad [7.1]$$

$$E = R/C \qquad\qquad [7.2]$$

where P is profit, E is economic efficiency, R and C are returns and costs, per unit production. As stated previously, some studies have also used the inverse of economic efficiency to estimate economic values. The reasons for this will be explained below. One advantage of economic efficiency, as compared to profit, that should already be apparent is that economic efficiency is independent of the units used to compute R and C. If in the previous chapter we showed that profit will be different if computed per unit product, per animal or per enterprise; this will not be the case for economic efficiency. Since the units of R and C will be the same, E is a unitless number. Thus on the basis of economic efficiency it is also possible to compare different species and production systems. Furthermore, since R and C will generally be approximately equal, E will generally be close to unity.

One important disadvantage of both these criteria is that both R and C will tend to vary over time, as discussed in Chapters 2 and 4. Thus "biological efficiency" (Dickerson, 1982) has been suggested as an alternative to economic efficiency. Biological efficiency is defined as unit output per unit feed energy input. Following the notation of Chapter 5, and assuming all quantities are measured on an enterprise basis, we can construct the following equations:

$$R = A_1(x_1) \qquad\qquad\qquad [7.3]$$

$$C = C_n + C_d F \qquad\qquad\qquad [7.4]$$

$$E_B = x_1/F = [C_d R]/[A_1(C - C_n)] \qquad\qquad\qquad [7.5]$$

Where A_1 is the price of a unit product, x_1 is quantity of product produced per enterprise, C_n are non-feed costs of the enterprise, C_d is the cost of a unit feed, F is the quantity of feed given, E_B is biological efficiency, and the other terms are as defined above. Note that the middle term of equation [7.5] is in terms of biological inputs and outputs, while the right-hand term is in terms of economic units. Since in many production systems, the main economic component of both feed and product is protein, biological efficiency can alternatively be defined in terms of input and output of protein, rather than gross feed energy and product. Although biological efficiency will be more constant over the long-term than either profit or economic efficiency, it is not a very useful criterion for economic evaluation. As pointed out by Dickerson (1982) it ignores the differing costs of feed for different species, and the differing value of products (e.g. protein vs. milk fat, or meat of old vs. young animals). In addition it is possible to increase economic efficiency without changing biological efficiency. For example breeding for disease resistance or calving ease may reduce non-feed costs and therefore economic efficiency without affecting biological efficiency.

A fourth criterion that can be considered is return on investment, I_w, defined as follows:

$$I_w = P/C_w = (R - C)/C_w \qquad [7.6]$$

where C_w is investment, and the other terms are as defined previously. Similar to efficiency, I_w will be a pure number. If all costs are included in C_w then I_w will be equal to $E - 1$. Moav (1973) suggested that costs that are a function of the quantity of production, but independent of weight and number of animals raised should not be included in C_w, specifically the feed required to produce the product. This is because decisions on the quantity of investment are taken before the product is produced. Thus Moav (1973) defined I_w as costs per unit weight of animal. This criterion is probably only of interest to a potential new investor, or an investor who is contemplating expansion. Similar to other profit-based criteria, I_w will be correct only for a given situation. Since the results of nearly all breeding decisions will be long-term, it is difficult to justify this criterion for economic evaluation of trait changes.

7.3 Economic evaluation of trait differences by economic efficiency

In the previous chapter, the economic values of unit changes in trait values were computed by taking the partial differentials of profit equations with respect to each trait. This method can also be applied to economic efficiency. This will be illustrated using the example of egg production given in equation [6.17]. Profit per hen, P_2 was computed as follows:

$$P_2 = (A_1 - F_1)x_1x_2 - K_2x_1 - K_3x_3 - K_4 \qquad [7.7]$$

where A_1 is income per gram egg; F_1 is fixed costs per gram egg; x_1 is number of eggs, x_2 is egg weight, x_3 is hen weight; and K_2, K_3, and K_4 are fixed costs per gram egg, per egg and per hen, respectively. In order to differentiate between costs and returns, K_1 was replaced by $A_1 - F_1$. The inverse of economic efficiency is now computed as follows:

$$E_i = \frac{F_1x_1x_2 + K_2x_1 + K_3x_3 + K_4}{A_1x_1x_2} \qquad [7.8]$$

The economic values of x_1, x_2, and x_3 can now be computed by taking the partial differentials of E_i with respect to these three traits. One reason that E_i has been preferred is that calculation of partial differentials will generally be easier for this function. These partial differentials will be equal to the partial differentials presented in the last row of Table 6.4, divided by A_1, which is a constant. Since, as explained in Chapters 3 and 6, multiplying the economic values by a constant will not affect the rate of genetic progress, we can conclude that, at least in this example, economic values by the criterion of economic efficiency, and by profit per unit product will be the same. This result is generally true and will now be explained in more detail.

7.4 Conditions for equality of economic values as computed for different profit criteria and economic efficiency

In the previous chapter, we showed that different economic values are obtained when profit is computed by different criteria, such as per dam, per progeny, or per unit product. Brascamp, Smith, and Guy (1985) and Smith, James, and Brascamp (1986) derived three conditions for equality of economic weights as computed by different profit criteria and by economic efficiency: 1) zero profit, 2) disregarding increased profit that can be achieved by rescaling of the enterprise, and 3) disregarding increased profit that can be obtained by correcting inefficiencies in the production system. Following their explanation, we will first use illustrative examples, and then prove the general principles.

We will start with the example of pig production from Chapter 6, under the condition of zero profit. The concept of "zero profit" was introduced in Section 8 of Chapter 2. We explained that over the long-term, the price for any commodity will tend to equal both the marginal and average cost of production. Thus if profit is computed as returns minus costs, profit will tend toward zero. We further explained that "zero profit" does not mean that the producer receives no compensation from production, but rather that a "reasonable profit", necessary to make production worthwhile, is included in the "costs" of production.

In equation [6.21] profit per pig, P_1 was computed as follows:

$$P_1 = K_1 - K_2x_2 - K_3/x_1 \qquad [7.9]$$

where x_1 is number of pigs weaned per sow, x_2 is slaughter age, K_1 is income per pig less costs independent of x_1 and x_2, K_2 is costs dependent on x_2, and K_3 is fixed costs per sow. Defining K as income per kg pig less costs per kg pig, and x_3 as slaughter weight we can rewrite this equation as follows:

$$P_1 = Kx_3 - K_2x_2 - K_3/x_1 \qquad [7.10]$$

as in the previous chapter we can compute profit per kg pig marketed, P_3, by dividing equation [7.10] by x_3 as follows:

$$P_3 = K - K_2x_2/x_3 - K_3/(x_1x_3) \qquad [7.11]$$

Similarly, profit per sow, P_2, can be computed by multiplying equation [7.10] by x_1:

$$P_2 = Kx_1x_3 - K_2x_1x_2 - K_3 \qquad [7.12]$$

The partial differentials of these three profit criteria with respect to x_1, x_2, and x_3 are given in the top three rows of Table 7.1. Under the assumption of $P_1 = P_2 = P_3 = 0$, these partial derivatives can be rewritten in the form appearing in the bottom three rows of Table 7.1.

Table 7.1. Partial differentials of profit with respect to pigs/sow (x_1), slaughter age(x_2) and slaughter weight (x_3).

Profit criteria	Partial derivatives		
	x_1	x_2	x_3
Per pig (P_1)	$\dfrac{K_3}{x_1^2}$	$-K_2$	K
Per sow (P_2)	$Kx_3 - K_2x_2$	$-K_2x_1$	Kx_1
Per kg pig (P_3)	$\dfrac{K_3}{x_1^2 x_3}$	$\dfrac{-K_2}{x_3}$	$\dfrac{K_2x_2}{x_3^2} + \dfrac{K_3}{x_1x_3^2}$
Per pig ($P_1 = 0$)	$\dfrac{K_3}{x_1^2}$	$-K_2$	K
Per sow ($P_2 = 0$)	$\dfrac{K_3}{x_1}$	$-K_2x_1$	Kx_1
Per kg pig ($P_3 = 0$)	$\dfrac{K_3}{x_1^2 x_3}$	$\dfrac{-K_2}{x_3}$	$\dfrac{K}{x_3}$

We see that the partial derivatives by each criterion are now proportional. That is, the partial derivatives of P_2 are equal to the corresponding partial derivatives of P_1, multiplied by x_1, while the partial derivatives of P_3 are equal to the partial derivatives of P_1, divided by x_3. Since, as shown in Chapters 3 and 6, it is the *ratios* of the economic values rather than their *absolute values* that determine the direction of selection, the economic values are now the same by all three criteria.

7.5 Proofs of conditions for equality of profit criteria

We will now present the general proof of Brascamp, Smith and Guy (1985). x will denote a vector of biological traits, and k_1, k_2, and k_3 will denote vectors of economic constants. Assume that profit is expressed as a function, $f(x,k_2)$, of x and k_2. We now define $h(x,k_1)$ as a different profit criterion computed by dividing $f(x,k_2)$ by a function $g(x,k_3)$. Algebraically:

$$P = h(x,k_1) = f(x,k_2)/g(x,k_3) \qquad [7.13]$$

In the example given above, $x = [x_1, x_2, x_3]$, $k_2 = [K, K_2, K_3]$, $P_3 = h(x,k_1)$, $P_1 = f(x,k_2)$, and $g(x,k_3) = x_3$. Computing the partial differentials of $h(x,k_1)$ with respect to x, we have:

$$\frac{\delta[h(x,k_1)]}{\delta x} = \frac{f(x,k_2)\delta[g^{-1}(x,k_3)]}{\delta x} + \frac{g^{-1}(x,k_3)\delta[f(x,k_2)]}{\delta x} \qquad [7.14]$$

If zero profit we have:

$$P_1 = f(x,k_2) = 0 \qquad [7.15]$$

$$\frac{\delta[h(x,k_1)]}{\delta x} = \frac{g^{-1}(x,k_3)\delta[f(x,k_2)]}{\delta x} \qquad [7.16]$$

Thus if $g^{-1}(x,k_3)$ is a constant, the economic values for x will be proportional by these two profit criteria. Even if $g(x,k_3)$ is a function of x, as in the case above, this can still be considered a constant if $g(x,k_3)$ is evaluated for the means of x.

Using the same line of reasoning, it can now be explained why economic values derived for profit per unit product will be proportional to economic values for the inverse of economic efficiency. Assume that profit is computed for some criterion other than unit product. This criterion can then be converted to profit per unit product by dividing by production in the units of the profit criterion as follows:

$$P_Q = (R - C)/x_1 = R/x_1 - C/x_1 \qquad [7.17]$$

where P_Q is profit per unit product, R and C are returns and costs for the alternative profit criteria, and x_1 is production. Note that R/x_1 is equal to returns per unit product, which is a constant. Therefore computing the partial differentials of P_Q with respect to x gives:

$$\frac{\delta P_Q}{\delta x} = \frac{-\delta(C/x_1)}{\delta x} = \frac{-A_1\delta(C/R)}{\delta x} \qquad [7.18]$$

Where A_1 is price per unit product. Thus the partial differentials for P_Q will be equal to the partial differentials for the inverse of economic efficiency, times $-A_1$, which is a constant.

The partial differentials for profit per unit cost, or profit per unit investment, I_W, can be computed as follows:

$$\frac{\delta I_W}{\delta x} = \frac{\delta[(R - C)/C]}{\delta x} = \frac{\delta(R/C)}{\delta x} \qquad [7.19]$$

which are the partial differentials for economic efficiency ("actual" economic efficiency, not the inverse of economic efficiency). Thus economic values computed for economic efficiency and its inverse will be equal at zero profit, but unequal for any other profit value.

We will now consider the second condition for equivalence, that is subtracting any gain in profit that could have been obtained by changing the scale of the enterprise. Consider profit per pig in equation [7.10]. Assume that slaughter age and number of pigs/sow are equal for all individuals. We can then simplify the equation as follows:

$$P_1 = Kx_3 - K_4 \qquad [7.20]$$

where $K_4 = K_2x_2 - K_3/x_1$. For an enterprise of m pigs, profit per enterprise is computed as follows:

$$P_4 = m(Kx_3 - K_4) \qquad [7.21]$$

If x_3 is increased by a small amount, denoted Δx_3 then the change in profit, ΔP_4, is computed as follows:

$$\Delta P_4 = mK\Delta x_3 = (P_4 + mK_4)\Delta x_3/x_3 \qquad [7.22]$$

Thus, as was illustrated in Chapter 6, the change in profit will also depend on the previous profit level and on the production costs. Original production was mx_3. An equal change in output could have been obtained by increasing m by Δm, computed as follows:

$$\Delta m/m = \Delta x_3/x_3 \qquad\qquad [7.23]$$

The change in profit due to rescaling, which we will denote ΔP_5 can then be computed as follows:

$$\Delta P_5 = P_4(\Delta m/m) = P_4(\Delta x_3/x_3) \qquad\qquad [7.24]$$

Thus the change in profit due to increase in x_3 after subtracting the gain in profit that could have been obtained by rescaling the enterprise is:

$$\Delta P_4 - \Delta P_5 = mK_4(\Delta x_3/x_3) \qquad\qquad [7.25]$$

Note that the right-hand side of equation [7.25] is now independent of P_4. Thus we see that the same result would be obtained for any original profit level. Since we proved above that all profit criteria will give proportional economic weights for zero profit, and since equation [7.25] is independent of the profit level, we can deduce that economic weights computed after correction for increased production will also be independent of the profit criteria.

In this example breeding increased production. Alternatively breeding could change the value of unit product, or reduce the costs of production. If profit is not equal to zero in the original state, then part of the change in profit due to change in either product value or cost of production could also be matched by rescaling the enterprise. It can readily be shown that for these cases as well, the economic values of the different traits will be independent of the profit criteria if the change in profit that could have been obtained by rescaling is subtracted.

We will now present the general proof of Smith, James, and Brascamp (1986) for this postulate. Assume that for a given profit equation there is a scaling factor, α, that produces proportional effects on costs and returns so that:

$$\frac{1}{R}\frac{\delta R}{\delta \alpha} = \frac{1}{C}\frac{\delta C}{\delta \alpha} \qquad\qquad [7.26]$$

A small change in a trait will then give a change in profit, ΔP_1, as follows:

$$\Delta P_1 = \Delta R - \Delta C = \left[\frac{\delta R}{\delta x} - \frac{\delta C}{\delta x}\right]\Delta x \qquad\qquad [7.27]$$

Assume now that the enterprise is rescaled by an amount $\Delta \alpha$ to match the change in output obtained by increasing x by Δx. That is:

$$\frac{\delta R}{\delta \alpha} \Delta \alpha = \frac{\delta R}{\delta x} \Delta x \qquad\qquad [7.28]$$

The change in profit due to rescaling, ΔP_2, is computed as follows:

$$\Delta P_2 = \left[\frac{\delta R}{\delta \alpha} - \frac{\delta C}{\delta \alpha} \right] \Delta \alpha \qquad\qquad [7.29]$$

By rearranging equation [7.26]:

$$\frac{\delta C}{\delta \alpha} = \frac{C}{R} \frac{\delta R}{\delta \alpha} \qquad\qquad [7.30]$$

Substituting equations [7.28] and [7.30] into [7.29] gives:

$$\Delta P_2 = \left[\frac{\delta R}{\delta x} - \frac{C}{R} \frac{\delta R}{\delta x} \right] \Delta x \qquad\qquad [7.31]$$

Then the net value of the genetic improvement, ΔP_3 can be computed as follows:

$$\Delta P_3 = \Delta P_1 - \Delta P_2 = \left[\frac{C}{R} \frac{\delta R}{\delta x} - \frac{\delta C}{\delta x} \right] \Delta x \qquad\qquad [7.32]$$

The differential of profit with respect to x after correction for the change of scale is then:

$$\frac{\delta P_3}{\delta x} = \frac{1}{R} \left[C \frac{\delta R}{\delta x} - R \frac{\delta C}{\delta x} \right] = \frac{C^2 \delta(R/C)}{R \delta x} = \frac{C \delta E}{E \delta x} \qquad [7.33]$$

Where $E = R/C$. $C/E = C^2/R$, which is a constant for all traits. Thus the relative economic weights will be proportional to $\delta E/\delta x$, which, as shown above, is independent of the criteria used to compute profit. The same result will be obtained if the enterprise is scaled to equal input or equal profit, the only difference being the proportionality constant. For equal input the proportionality constant will be C and for equal profit, the proportionality constant will be C(E − 1).

This line of reasoning was also extended by Smith, James, and Brascamp (1986) to cover the more general case of profit from differing rates of production and female reproduction, and can readily be extended to more complicated profit equations.

7.6 Conclusions on the choice of criteria to compute economic values

In Chapter 4 we first considered the question of the appropriate criteria for economic evaluation of genetic differences. Although all of the criteria suggested have advantages and disadvantages, those considered most appropriate were profit and the inverse of economic efficiency. In Chapter 6 we were confronted with the anomaly of Moav (1973) that the relative economic values of different traits are different for different profit criteria. This means that different entities involved in breeding could have different objectives. Accepting the conditions for equality presented above, this problem can now be considered resolved.

Before completing this discussion, we should note though, that in practice rescaling is often not a viable option. A farm may be organized to handle a set number of cows. Thus, even if the enterprise is in a positive profit situation, it may not be possible to increase the scale of the operation, even if increasing production per cow would increase both returns and profit. However, these considerations are generally only short-term considerations for individual producers. Thus over all producers, or for the national interest, it would seem that economic values should be computed either for economic efficiency, its inverse, or a profit criterion which is unaffected by scale, and these criterion will result in proportionate economic values for all traits under selection.

7.7 "Empirical" methods for estimating economic values

The methods presented above to estimate the economic values of different traits assume that the simple equations presented are basically correct and that the economic constants in these equations can be accurately estimated. Often in practice neither of these assumptions is correct. An alternative method was suggested by Dickerson (1982) for economic efficiency, and applied by Sivarajasingam *et al.* (1984) for profit per cow in a dairy herd.

Sivaragasingam *et al.* (1984) used Linear Programming (LP) to compute the expected profit of a progeny of each of 71 dairy sires, defined as an average daughter's expected returns over variable costs attributable to the sires' estimated transmitting abilities for the traits under consideration. An index of total economic merit was computed as the amount by which optimal LP net profit would be reduced by mating a cow to a given sire, instead of the most profitable sire. These authors state that the "true" economic relationships of the component traits are the "partial correlations" of the component traits on the index of total economic merit. Partial correlations are defined as the correlation of the regressor and the response if all other regressors are held constant. Since the goal of selection index is the maximization of genetic progress, the appropriate economic values for use in selection index are the partial regressions of each trait on the index of economic merit. They also computed the partial regression coefficients and the coefficient of determination for the index.

Dickerson (1982) suggested computing the partial regression coefficients of the individual traits on economic efficiency from a simulation model. The advantages of these "empirical" methods for computing economic trait values are: 1) the economic values will be by definition linear functions of the trait values, and 2) it is possible to include factors and relationships that may not be readily included in profit equations. The disadvantages are: 1) lack of generality, as these methods are applicable only to the sample populations measured or simulated, 2) they do not account for changes in relative economic values due to selection, which is, in fact, the reason why the economic values of some traits are nonlinear, and 3) they can only be applied if an appropriate data sample is available, or if the parameter values of the simulation are known. Thus, in conclusion, it does not seem that these alternative methods can be recommended over the analytical methods described previously.

7.8 Summary

In this chapter we concluded the discussion of methods for estimating the economic values of the component traits in the selection index. In the previous chapter we were presented with the dual dilemmas: 1) the economic values of most traits will not be linear functions of the trait values, and 2) different relative trait values are obtained depending on the profit criteria. In this chapter we considered economic efficiency, biological efficiency, and return on investment as alternatives to profit as criteria for computing economic values. Of these alternatives, we showed that only economic efficiency is generally applicable. We explained how Smith, James, and Brascamp (1986) were able to resolve the second dilemma for most actual conditions of interest. Under these conditions, the economic values for the component traits computed for profit by any criteria and economic efficiency will be proportional. Finally we briefly discussed "empirical" methods for computing economic values based on performance records or simulated data.

Chapter Eight

Long-term Considerations

8.1 Introduction

Animal breeding is by its nature a long-term process. For example, some results of the most important breeding decisions in dairy cattle are only realized after ten years. Thus a number of considerations that may not be important for relatively short-term processes are of major importance for most animal breeding programs. Furthermore, the different costs and returns in animal breeding procedures are realized at different times, and with differing probabilities. Thus, factors that affect costs and returns over the long-term must be considered in the economic evaluations of genetic differences. Long-term considerations will affect both the attractiveness of investment in breeding programs, and the relative economic values of the individual traits included in the selection index. The main long-term considerations of animal breeding programs are the discount rate, risk, profit horizon, and reproduction rates. Discount rate and profit horizon were discussed in general terms in Chapter 2. In this chapter we will discuss these factors in more detail. In the final section of this chapter before the summary, we will give concrete examples of computation of economic values over the long-term.

8.2 Discounting of costs and returns

The first consideration with respect to discounting of costs and returns is which discount rate is appropriate. Most studies that have discounted costs and returns in animal breeding programs have used rates of 5 to 15%. Smith (1978) lists three alternative criteria for setting discount rates in breeding programs. First, d_s, the social time preference rate. This is the lowest rate and is appropriate for minimal risk investments in the national interest, such as building roads, ports, or public buildings. Second, the opportunity cost rate, which is the cost of borrowing in the financial market. The third alternative is a synthetic rate which allows for the returns forgone by diverting capital from the higher return rate to the lower d_s rate, but discounts the returns forgone and the actual returns by the d_s rate. The main causes for divergence between the two rates are due to the effects of inflation, risk, and taxes on private investment.

As explained in Section 2.9, "nominal" interest rates are strongly affected by the inflation rate. In equation [2.9] we showed how the discounted value of current investment is also affected by the inflation rate. The situation presented in Chapter 2, an initial investment and equal monetary returns year after year, is not typical of breeding programs because inflation will affect the nominal values of both costs and returns. Thus for breeding programs it is necessary to correct the nominal interest rate, d_i, by the rate of inflation, d_t as follows:

$$1 + d_i = (1 + d_q)(1 + d_t) \qquad [8.1]$$

$$d_q = (d_i - d_t)/(1 + d_t) \qquad [8.2]$$

where d_q is the "real" interest rate corrected for inflation. For moderate rates of inflation d_q can be computed approximately as $d_i - d_t$. Although nominal interest rates have varied greatly over the last century, the real interest rate has remained quite stable over the long-term at close to 3% (Smith, 1978).

In addition to inflation, risk and taxation should also be included in the required nominal rate of return. Considering these factors the required nominal rate of return, d_r, can be computed as follows:

$$d_r = \frac{(1 + d_q)(1 + d_t)/(1 - d_k) - 1}{(1 - d_x)} \qquad [8.3]$$

where d_k is the risk, d_x is the tax rate, and the other terms are as defined above. Clearly the nominal rate can be considerably higher than d_q, even for relatively low rates of inflation, risk and taxation. For example, if $d_q = 0.04$, $d_t = 0.05$, $d_k = 0.02$, and $d_x = 0.1$; then $d_r = 0.127$, or 12.7%. This rate is similar to current nominal interest rates in most developed countries with moderate inflation rates, but considerably higher than d_q, which should approximate the d_s rate discussed above.

8.3 Estimating discounted returns and costs for a single trait with discrete generations

Returns from breeding programs, unlike nearly all other investments, are *cumulative*. This important distinction will be elaborated with an example. A company invests in a new piece of machinery, which increases the efficiency of production, and therefore the net income of the enterprise. Eventually though the machine will either be discarded or replaced. Therefore this investment will generate additional income for a finite period. Now we will compare this example to the situation in genetic improvement. Assume that milk production per cow is increased genetically by 100 kg. Once this genetic gain has been achieved in the population, no additional investment is necessary to maintain it, and, contrary to the previous example, this gain will never "wear out" or need to be "replaced." Unless returns are discounted, the same gain in profit is obtained year after year, and the total gain from any amount of genetic improvement will tend to infinity. If returns are discounted, the gain of a single cycle of genetic improvement extended to infinity can be computed from equation [2.8]. This equation gives the minimum acceptable annual return, V, from an initial investment of N with a discount rate of d_i. We will repeat this equation:

$$Nd_i = V \qquad\qquad\qquad [8.4]$$

If V is now taken as the value of one year of genetic improvement, then the cumulative return (R), extended to infinity will be:

$$R = V/d_i \qquad\qquad\qquad [8.5]$$

Note that R, return from the breeding program, has replaced N of equation [8.4]. Continuing with the example given in Chapter 2, if the annual value of a cycle of genetic improvement is $10, and the discount rate is 0.1, than the discounted value of this gain, year after year, to infinity is $100. Alternatively, equation [8.5] can be derived as follows: The return from a breeding program to infinity, will be equal to the sum of a geometric progression of the form $V(r^1 + r^2 + \ldots + r^n + \ldots + r^\infty)$, where $r = 1/(1 + d_i)$, and n is the number of years from the beginning of the program. The sum, S, of a standard geometric progression of the form Vr^n from $n = 0$ to $n = T - 1$ is computed as follows:

$$S = \sum_{n=0}^{T-1} Vr^n = V(1 - r^T)/(1 - r) \qquad\qquad [8.6]$$

Thus in our case, the net return is computed as S with $T = \infty$ less V, as follows:

$$R = \sum_{n=0}^{\infty} Vr^n - V = \frac{V}{1 - r} - V = \frac{Vr}{1 - r} = V/d_i \qquad [8.7]$$

Generally there will be a lag period of several years until the first realization of any gain from genetic improvement. This will require a further discounting of returns as follows:

$$R = \sum_{t=0}^{\infty} Vr^n - \sum_{t=0}^{t-1} Vr^n = \frac{V}{1-r} - \frac{V(1-r^t)}{1-r} = \frac{Vr^t}{1-r} = \frac{V}{d_i(1+d_i)^{t-1}} \qquad [8.8]$$

where t is the number of years until the first return is realized.

In Chapter 2 we discussed the concept of "profit horizon". Estimating all returns to infinity is not realistic. Certainly no economic enterprise, and not even a government, will make decisions now based on returns expected 100 years in the future. A more realistic alternative is to estimate returns and costs for a given time period, say twenty years, under the assumption that all returns accruing after the profit horizon have a current value of zero. Cumulative return will then be equal to the sum of a geometric progression of the form $V(r^t + r^{t+1} + \ldots + r^T)$, where T is the profit horizon in years. The net return is then computed as the difference of two progressions, as follows:

$$R = \sum_{t=0}^{T} Vr^n - \sum_{t=0}^{t-1} Vr^n = \frac{V(1-r^{T+1})}{1-r} - \frac{V(1-r^t)}{1-r} = \frac{Vr^t(1-r^{T-t+1})}{1-r} \qquad [8.9]$$

Substituting $r = 1/(1 + d)$ gives (Smith, 1978):

$$R = \frac{V}{d_i(1+d_i)^{t-1}} [1 - 1/(1+d_i)^{T-t+1}] \qquad [8.10]$$

As T approaches infinity, the term in square brackets approaches unity, and equation [8.10] becomes equal to equation [8.8]. The term in brackets can be used to estimate the proportion of the total returns of a cycle of genetic improvement for a given discount rate and lag time. For example the "half-life" of a cycle of genetic improvement can be calculated by setting this term equal to 0.5 and solving for T with known values for y and d_i. For example if $d_i = 0.03$ and $y = 5$ years, then the half-life of a cycle of genetic improvement will be 27 years. Thus if the profit horizon is set at 20 years, less than half of the total gain will be realized within the profit horizon with this particular interest rate. Conversely if $d_i = 0.1$, the half-life will be 13.3 years with the same lag period, and the 90% life will be 30 years. For the simple case of $t = 1$ and $T = $ infinity, the ratio of expected gain for two different interest rates and equal genetic gain can be computed as follows:

$$R_1/R_2 = d_2/d_1 \qquad [8.11]$$

where R_1 and R_2 are the expected cumulative gains with interest rates of d_1 and d_2, respectively. In conclusion, for relatively low discount rates, determination of the profit horizon can have a major effect on the expected total gain; while for relatively high discount rates, the difference between a finite and infinite profit horizon will be minimal.

We can now extend this equation to consider an ongoing breeding program with a genetic gain of V each year. The *cumulative* discounted return can then be computed as the sum of a progression of the form $V[r^t + 2r^{t+1} + \ldots + (T - t + 1)r^T]$. The sum of this progression is computed as follows (Hill, 1971):

$$R = V \left[\frac{r^t - r^{T+1}}{(1-r)^2} - \frac{(T - t + 1)r^{T+1}}{1 - r} \right] \tag{8.12}$$

For a discount rate of 0.08, a profit horizon of 20 years, and first returns after 5 years, R = 32.58V. For an infinite profit horizon, equation [8.12] reduces to:

$$R = \frac{Vr^t}{(1 - r)^2} = \frac{V}{d^2(1 + d)^{t-2}} \tag{8.13}$$

Continuing the previous example of a discount rate of 0.08, and y = 5, for an infinite profit horizon, R = 124.04V. Thus even with a relatively high discount rate, a little bit of genetic improvement goes a long way.

Until now we have been considering additive genetic improvement, which is generally considered to be cumulative. Not all genetic improvement is additive, and therefore cumulative. In many domestic species the commercial animal is a crossbreed produced by breeding different lines. Since it is necessary to reproduce the crossbreed each generation, any gain in efficiency specific to the crossbreed will not be additive. Thus for crossbreeding we can set T equal to the generation interval. Therefore with low discount rates, it is much more profitable to utilize additive genetic variance than heterosis. For example, if t = 1 and d_i = 0.03 the net present value of one unit of genetic gain extended to infinity will be 1/0.03 = 33. To obtain the same discounted value by crossbreeding with T = 1, requires that V/1.03 = 33, or a nominal gain of V = 34 units. This is, of course, an extreme example, but even with T = 10, and d = 0.15, the nominal gain from crossbreeding must be six-fold the nominal additive gain, so that the current discounted values extended to the profit horizon will be equal.

We will now briefly consider the net present value of the costs of a breeding program, under the assumption of constant costs per year. Unlike genetic gain, costs of a breeding program are not cumulative. With an infinite profit horizon, equation [8.5] can be used to compute the costs with V replaced with C_c, the annual costs of the breeding program. For a finite profit horizon, equation [8.9] can be used to compute the net present value of the costs, C. With first costs in the following year, C is computed as follows (Hill, 1971):

$$C = \frac{C_o r(1 - r^T)}{1 - r}$$ [8.14]

Using the values of $T = 20$, $d = 0.08$, and $r = 0.926$; the net present value of the costs of the breeding program will be $9.82C_o$. Thus for $t = 5$, net profit will be positive if $V > 0.31C_o$. Note that profit can be positive even if yearly costs are greater than the revenue from yearly genetic gain. Again, this is due to the fact that genetic gains are *cumulative*, while costs are not. Extended to an infinite profit horizon, $C = 12.5C_o$. As computed above for this case $R = 124V$. Thus profit will be positive if $V > 0.1C_o$. Examples will be considered in Chapter 12.

8.4 Dissemination of genetic gain in populations for a single trait

In the previous section we considered the net present value of genetic selection on a single trait expressed once per generation with discrete generations. In this section we will expand the calculations of the previous section to a situation of overlapping generations and multiple trait expressions per individual. The multitrait situation will be considered in detail in the following section. In Chapter 1 we defined the four paths of genetic inheritance, sire to sire, sire to dam, dam to sire, and dam to dam. We showed that annual genetic gain in the population can be computed by equation [1.25], that is, the sum of the genetic gains per generation by the four paths of inheritance, divided by the sum of the four generation intervals. However this equation will be correct only for a breeding program at equilibrium. If a new program is started, or if an existing program is modified, there will be a lag before any gain is obtained, and then genetic gains will fluctuate around the equilibrium value for several generations (Owen, 1975). Equations [8.5] through [8.13] are correct, only under the assumption that the rates of genetic gain and generation intervals are the same along the four paths of inheritance. Generally this is not the case. Fertility rates, and therefore possibilities for selection are generally greater for males, while many important traits, related to female reproduction, are expressed only in females. Due to both biological and breeding considerations, generation intervals are also different along the four paths. Finally, the time and frequency of trait expression can vary.

These last considerations will be explained with the example of dairy cattle. The main traits under selection are related to milk production. To evaluate the net present value of a sire's semen for milk production, we must first consider the probability that an insemination from this sire will result in a milk-producing daughter. If a milk-producing daughter results, this cow can have several lactations. It is necessary to account both for the probability that a given lactation will occur, and the differing time-lag from the initial investment to realization. Finally the daughter will have a variable number of offsprings, each of which will receive only half of the genetic complement passed to the original daughter.

If we wish to compare the net present value of genetic improvement for meat production from the dairy herd, we are faced with an entirely different situation. Generally calves will be slaughtered at the age of one year. Thus the gain from increased meat production will be realized sooner, but will of course be realized only once. Furthermore, no gain will be accrued in future generations from these individuals, since they will invariably be slaughtered prior to mating. Thus increasing slaughter rate increases the probability of the realization of this trait in the short-term, but decreases the rate of genetic dissemination in the long-term.

A number of studies have addressed various aspects of these problems. As in the previous chapters we will progress from simpler, more specific cases to the more complex, general cases. We will first consider the economic evaluation of the genotype of a single individual for a single trait, with a single expression per animal, such as meat production (McClintock and Cunningham, 1974). They called their method the "discounted gene flow technique". Representations are simplified by the use of matrix algebra. Extending the calculations of the previous section, the net present value of the unit semen from a given sire, N, for a single trait can be computed as follows:

$$N = d'u(BV)a \qquad\qquad [8.15]$$

where u is a column vector and d' is a row vector both of dimension equal to the number of years from insemination to profit horizon; BV, a scalar, is the sire's breeding value for the trait in question; and a, also a scalar, is the economic value of a unit change in the trait. The elements of u represent the expectation of the fraction of the sire's genotype that will be expressed in his progeny in a given year. The elements of u are computed by multiplying the probability of the trait expression in a given year by the fraction of the genome of the original sire passed to each descendant, and summing over all possible descendants that could express the trait in that particular year. The elements of d are the appropriate discounting factors for the elements in u. The jth element of d' is computed as follows:

$$d_j = 1/(1 + d_i)^k \qquad\qquad [8.16]$$

where k is the time period in years from original investment to mean trait expression, and d_i is the discount rate. Although there is no general formula for computing the elements of **u**, McClintock and Cunningham (1974) provided formulas to compute this vector for the specific situation in their study. In the above discussion we have assumed that the trait in question is expressed once yearly. If this is not the case, then the dimension of **u** will be the number of different times that the trait can be expressed up to the profit horizon over all possible descendants of the original sire, and k must be computed accordingly.

For a trait that can be expressed several times by each individual, such as milk or wool production, equation [8.15] can be expanded as follows (McGilliard, 1978):

$$N = \mathbf{d'Um}(BV)a \qquad\qquad [8.17]$$

where **m** is a column vector of length equal to the possible number of expressions of the trait (lactations), **U** is a year-by-parity matrix, and the other terms are as defined previously. If all expressions of the trait have equal value then **m** will be a column of ones. If, as in the case of milk production, lactation yield increases with parity, then one element of **m** will have the value of unity, and the other elements will have values in proportion to the "standard" trait expression. The breeding value will be estimated relative to the "standard" trait expression. The elements of **U** are computed by multiplying the probability of the trait expression in a given year-lactation combination, times the fraction of the genome of the original sire passed for each descendant, and summing over all possible descendants that could express the trait in that particular year-parity combination. As in the previous example there is no general formula for computing **U**, but McGilliard (1978) provided an algorithm for computing this matrix for the specific example of dairy cattle.

Equations [8.15] and equations [8.17] can readily be expanded to deal with the multitrait situation. If several traits are expressed jointly, for example milk, butterfat, and milk protein production, then it is only necessary to replace (BV)a with the aggregate genotype, $H = \mathbf{y'a}$, also a scalar, as computed in Section 3.4. If the different traits are expressed at different times, and with differing probabilities, it will be necessary to compute $\mathbf{d'u}$ or $\mathbf{d'Um}$ for each trait. We are now confronted with the rather undesirable result that, unless all traits are expressed jointly, the relative economic values of the different traits will depend both on the discount rate and the profit horizon.

In this section we have removed the restraints of the previous section, but have so far only considered the net present genetic value of a single individual. We will now extend these calculations to evaluate the net present value of selection for a complete population with differing rates of male and female selection. General formulas for single-trait selection were derived by Hill (1974). As we have already noted in the previous examples, generations for most domestic animals overlap. For example both a cow and her daughter may be producing milk at the same time. Thus, although both records must be discounted equally, the expected genetic gains will be different. We will use Hill's example of pig production to derive the general formula for estimating genetic gain in a population with differing rates of male and female selection, unequal generation length, and overlapping generations.

Assume that sows farrow twice, at the age of one and 1.5 years, while boars are mated only once, and produce progeny at the age of one year. It will then be convenient to measure time in half-year units. The genetic makeup of the population can then be described by considering five groups of animals, males of age six months (one time unit), males of age one year, and females of ages six months, one year and 1.5 years. The passage of genes from one time unit to the next can then be described in matrix algebra by the following equations:

$$y_t = Zy_{t-1} \qquad\qquad [8.18]$$

where y_t is vector of the breeding values of these five groups of animals at time t and Z is a matrix that describes the passage of genes across these five groups of animals from time t − 1 to time t. Each element, z_{ij}, is computed as the proportion of genes in animals of sex-age class i at time t coming from animals of sex-age class j at time t − 1. For this specific example, Z is computed as follows:

$$
Z = \left[
\begin{array}{cc|ccc}
0 & 1/2 & 0 & 1/4 & 1/4 \\
1 & 0 & 0 & 0 & 0 \\
\hline
0 & 1/2 & 0 & 1/4 & 1/4 \\
0 & 0 & 1 & 0 & 0 \\
0 & 0 & 0 & 1 & 0
\end{array}
\right] \qquad [8.19]
$$

with the four blocks of this matrix corresponding to the pathways sires to sires, sires to dams, dams to sires, and dams to dams. Thus the genetic composition of a population after any number of generations, starting from any original population makeup, can be computed by successive applications of equations [8.18].

The matrix Z includes both the effect of ageing (represented by elements equal to unity) and reproduction (represented by positive elements less than unity). Only those elements due to reproduction are of interest for estimating genetic advancement. To remove the effect of ageing we will define a new matrix, Q, that includes changes in the population structure due solely to ageing. In the example given above, Q is computed as follows:

$$Q = \begin{bmatrix} 0 & 0 & 0 & 0 & 0 \\ 1 & 0 & 0 & 0 & 0 \\ \hline 0 & 0 & 0 & 0 & 0 \\ 0 & 0 & 1 & 0 & 0 \\ 0 & 0 & 0 & 1 & 0 \end{bmatrix} \qquad [8.20]$$

We can now compute the response to one cycle of selection at time t, r_t as:

$$r_t = (Z^t - Q^t)(y_{mo}G_m + y_{fo}G_t) \qquad [8.21]$$

where Z^t and Q^t are Z and Q to the tth power, y_{mo} and y_{fo} are the vectors that represent the original genetic makeup of the population for males and females, respectively, and G_m and G_t are scalars that represent the genetic superiority of the males and females selected as parents for the next generation. The dimension of the vectors r_t, y_{mo} and y_{fo} will be the sum of the number of male and female age groups equal to or below mating age. The non-discounted value of the response to one cycle of selection at time t, V_t, can then be computed as follows:

$$V_t = a'r_t \qquad [8.22]$$

where a is the vector of the economic values of a unit change in the trait for each group of animals. Defining v as a vector of the economic values of the gains in selection from time 1 to time t, the discounted gain in selection to time t, N_t, can then be computed as:

$$N_t = d_{it}'v \qquad [8.23]$$

where d_{it}' is a row vector of discount factors as defined above.

If the same selection procedure is performed on each successive group of animals born, then the response at time t from selection at time 1 is equal to response at time $t - 1$ to selection at time 0, and so forth. Then the total response to a continuous program of selection, R_t, starting from any base population structure can be computed as:

$$\mathbf{R}_t = \mathbf{r}_t + \mathbf{r}_{t-1} + ... + \mathbf{r}_o$$

$$= [(\mathbf{I} + \mathbf{Z} + \mathbf{Z}^2 + ... + \mathbf{Z}^t) - (\mathbf{I} + \mathbf{Q} + \mathbf{Q}^2 + ... \mathbf{Q}^t)]\mathbf{s} \qquad [8.24]$$

where $\mathbf{I}$ is the identity matrix, and $\mathbf{s}$ is computed as follows:

$$\mathbf{s} = \mathbf{y}_{mo}\mathbf{G}_m + \mathbf{y}_{fo}\mathbf{G}_f \qquad [8.25]$$

Replacing $\mathbf{r}_t$ with $\mathbf{R}_t$ in equations [8.22] and [8.23] gives the total discounted returns from a program of continuous selection up to time t.

These equations will become quite complex for most realistic population structures. It is likely that for most situations a reasonable approximation of the true economic values can be obtained by the equations of the previous section, which assume a constant rate of genetic gain per year.

8.5   Index selection for both current and future generation gains

We will now consider construction of the optimum selection index accounting for discounting of future gains and a limited profit horizon. We will consider three situations. The simplest situation is when all traits that are included in the selection index are realized at the same time, and with the same probability, for example milk, butterfat, and milk protein production. In this case, the effects of discounting future gains and a finite profit horizon will be the same for all traits, and the effects of these factors can be ignored in the computation of the economic values.

The second situation is when different traits are realized at different times and with possibly differing probabilities, but we are still interested only in the rate of genetic gain of the population. For example in swine, the number of progeny per dam, and growth rate are expressed at different times. Furthermore nearly all animals are slaughtered, but not all animals are used for reproduction.

In this case the equations of the previous section and Chapter 3 can be used to determine the optimum selection index. In Chapter 3 we defined the optimum selection index, $\mathbf{b'x}$, as the index with the maximum correlation, or minimum mean squared deviation from, the aggregate genotype $\mathbf{a'y}$. The vector of index coefficients, $\mathbf{b}$, was derived in equation [3.17] as follows:

$$\mathbf{b} = \mathbf{P}^{-1}\mathbf{Ca} \qquad [8.26]$$

Where $\mathbf{P}$ is the phenotypic variance matrix, $\mathbf{C}$ is the genetic covariance matrix between the traits in the selection index and the aggregate genotype, and $\mathbf{a}$ is the vector of economic values. We will now define the aggregate genotype of animals age-sex class j at time t, H_{jt}, as follows:

$$H_{jt} = a_{jt}'y \qquad [8.27]$$

where a_{jt} is the vector of non-discounted economic values of the traits included in the aggregate genotype at time t for the jth class of animals. The aggregate genotype over all classes of animals at time t can then be computed by expanding a_{jt}' into a matrix A in which each column represents the economic values for the jth class of animals at time t, and multiplying A by the proportion of genes in animals of each group at time t derived by reproduction from the group selected at year zero. This vector can be derived from equation [8.21] of the previous section as follows:

$$m_t = (Z^t - Q^t)m_o \qquad [8.28]$$

where m_t now represents the proportion of genes at time t in all age-sex classes derived from the original population distribution, m_o. The present economic value can then be computed by multiplying by the discount factor at time t. To account for all time periods from zero to t, the vector, m_t, is expanded into a matrix, M, in which each column represents a time period from 1 to t, and each row represents an age-sex class. This matrix is then multiplied by the vector of discount factors, d_{it}. Combining all elements we obtain the following equation:

$$b = P^{-1}CAMd_{it} \qquad [8.29]$$

with all terms as defined above. Note that the vector a of equation [8.26] is now replaced with AMd_{it}, which, after the appropriate matrix multiplication, will result in a vector of dimension equal to the number of traits in a.

The third situation we will consider is when the breeder or producer wishes to optimize jointly current production, gains in the current generation, and long-term genetic gains. This situation was considered by James (1978) for the case of meat and wool production from sheep. The producer should optimize jointly for current production of meat from culled animals, future wool production of ewes kept in the flock, descendants sold prior to mating for meat, and wool and meat produced by descendants used as parents for future generations. Of course a similar situation will exist for dual purpose, meat and dairy, cattle. This situation will also apply for calving traits in dairy cattle, even if meat production is of negligible economic value. In many countries dairy sires are evaluated for both their fertility and the calving ease of their daughters. Even though data on fertility will be collected on the daughters, this trait can be considered a phenotypic expression of the sire. Since the genetic correlation between male and female fertility is not significant (Ron, Bar-Anan, and Wiggans, 1984), the genetic component for male fertility passed to daughters will be expressed only if a daughter is used as a bull dam. Similarly, the main goal of selection for dystocia will be to reduce calving difficulty in the current generation of a sire's progeny. Since the genetic correlation between dystocia as a trait of the sire, and as a trait of the dam is low (Meijering, 1984; Thompson, Freeman, and Berger, 1981), selection of "easy calving" sires will not necessarily result in "easy calving" daughters.

The economically optimum selection index for this situation is computed by the following equation:

$$b = P^{-1}(P_C w + P_R A S d_{tt} + C A M d_{tt}) \qquad\qquad [8.30]$$

Where P_C is the phenotypic covariance matrix between the traits in the selection index and the traits in the economic objective; w is the vector of economic values of the index traits for animals in the current generation; P_R is the phenotypic covariance matrix between the trait values in the subsequent generation and the economic objective; S is the matrix, in which each column is the s-vector of equation [8.25] for a different t-value; and the other terms are as described previously.

Thus the overall index is made up of three subindices: the first for optimizing current returns, the second for optimizing gain in the next generation, and the third for optimizing long-term genetic gains. This result is analogous to the result of Henderson (1963) that selection index is a weighted sum of subindices, each of which would maximize gain in a single trait. If all traits included in the economic objective are also included in the index, then $P = P_C$, and equation [8.30] can be simplified accordingly.

8.6 Examples of computations of economic weights including discount rate and profit horizon

In this final section before the summary we will review three studies that have attempted to compute long-term economic values of different traits. The first example we will consider is Soller and Bar-Anan (1973). They compared the economic values of milk and meat production from a commercial dairy population. Their calculations were based only on a single generation of selection. Thus they considered differential discounting of the two traits, and differential probability of trait realization, but did not consider the possibility of differential gene flow for the two traits. This would be correct only if all animals were selected on the same criteria. In practice however, only bulls will be selected on an index based on both traits, while cows will be selected chiefly on milk production. They did consider three market situations: subsidized milk and meat prices and no production limits, subsidized milk prices and milk production quotas, and free market prices for both products. Over these three market conditions, the range in the relative economic values of kg live weight vs. kg milk production was from 7.2:1 to 8.2:1. Thus the ratio of economic values was quite robust to dramatic changes in market conditions.

Cunningham and Ryan (1975) studied the effect of variation in the discount rate and profit horizon for the same traits considered in the previous study, but under the assumption that a fraction of the dairy cows are bred to sires of a beef strain. They then assume that no progeny from these matings are allowed to reproduce. With a discount rate of 0.08, 98% of all economic benefit from an insemination will be realized within 15 years. Thus longer profit horizons need not be considered. Alternatively, returns computed to infinity will be nearly equal to returns computed to a profit horizon of 15 years. These results are different from those of Smith (1978), presented above in Section 8.3, who found that only 90% of the total net present gain would be realized after 30 years with a discount rate of 0.1. The discrepancy is due to the fact that Smith (1978) considered the situation of raising the genetic level of the entire population, while Cunningham and Ryan (1975) considered the consequence of a single insemination.

Although these authors considered the effect of varying the discount rate on the net present value of each trait, they did not consider the effect of the discount rate on the ratio of the economic values. Increasing the discount rate lowered the net present value of milk production slightly more than the net present value of beef, but the exact ratio can not be computed from the results as presented. Each 1% increase in the discount rate reduced the net present values by approximately 5% for milk and 4% for beef. Thus one can deduce that over the range of discount rates studied, 0.08 to 0.16, the ratio of the economic values would change only marginally.

Weller, Norman, and Wiggans (1984a) considered first, second, and third parity milk production as three separate traits, and estimated the relative economic values of milk production for each parity. This study assumed that lactations of parities greater than three would have a higher genetic correlation with third parity production, as opposed to first parity production. It should be now be clear that increasing the discount rate and reducing the profit horizon will increase the relative economic value of first, as opposed to third parity. As the profit horizon increased from 7 to 10 years, the relative economic values of first and third parity changed from approximately 2:1 to 1:2. However, after 10 years, the effect of further increases in the profit horizon was minimal. As the discount rate increased from 0.05 to 0.15 this ratio increased from 0.8:1 to 1:1. Varying the survival rate from parity to parity over the range of 0.7 to 0.9 had a greater effect on the relative economic values than did varying the discount rate.

In conclusion, all of these studies indicate that, in general, the relative economic values of traits are robust to realistic changes in the profit horizon, the discount rate, and the probability of income realization.

8.7 Summary

Since the different costs and returns in animal breeding procedures are realized at different times, and with differing probabilities, factors that affect costs and returns over the long-term must be considered in the economic evaluations of genetic differences. The main long-term considerations of animal breeding programs are the discount rate, risk, profit horizon, gene flow and commercial population size. Formulas were presented to account for the effects of these factors on the net present value of genetic selection. A major problem in including these factors in the computation of economic values is that they are often unknown or subjective. Most studies that have included discounting factors in the computation of economic trait values have shown that these factors can have major effects on the *absolute* economic values, but they will tend to have only minimal effects on the *relative* economic values of the traits considered. Thus, although the discount rate and profit horizon will have a major effect on the attractiveness of investment in breeding programs as compared to other investment alternatives, selection index coefficients will be robust to "reasonable" variation in these factors.

PART III

CONSTRUCTION, USES AND PROBLEMS OF MULTITRAIT SELECTION INDICES

In Parts I and II we covered the basic concepts necessary to study the economic aspects of animal breeding, and the methods for economic evaluation of genetic differences. Linear selection index was introduced in Chapter 3, and its basic properties were discussed. Application of selection index requires that the genetic parameters of the economic traits, and their economic values be known. Although it is relatively easy to derive reliable estimates of heritability, it is much more difficult to derive reliable estimates for genetic and environmental correlations. In fact many published reports have presented estimates of covariance components outside the parameter space. (Methods for estimation of genetic parameters have been dealt with very extensively in the literature and are not considered within the scope of this book. Although a great many references could be presented, we will mention only Henderson (1984), which is to date the most complete discussion of this topic.)

Even if good estimates of the genetic parameters have been derived, linear selection index can only be applied if the economic values of the different traits are linear functions of the trait values. In Part 2 we demonstrated that this is rarely the case. Not only are the economic values not linear functions of the trait values, they can also be different by different economic criteria (profit vs. economic efficiency), and will depend on extraneous factors, some of which are subjective, such as the profit horizon, or acceptable rate of risk.

The difficulties with the direct application of linear selection index have led to the development of alternative selection criteria. These alternatives will be considered in Chapter 9. In Chapter 10 we will use theoretical and simulation studies to compare linear and nonlinear selection indices.

Chapter Nine

Selection Indices for Nonlinear Profit Functions

9.1 Introduction

Although several alternatives to linear selection index have been considered, none is a complete solution to the difficulties encountered with linear selection index. The three main alternatives that have been suggested are nonlinear (quadratic, cubic, etc.) selection index, restricted selection indices, and linear indices derived by graphic methods. In this chapter we present the mathematical derivations of each index using matrix algebra, and then consider the advantages and disadvantages of each method. Where applicable, we will also present examples from the literature.

9.2 Quadratic models for the aggregate genotype

The concept behind quadratic aggregate genotype can be understood by the parallel to linear models. In a simple linear model, the dependent variable is assumed to be a linear function of the independent variables. If the linear model does not satisfactorily "fit" the data, it can be improved by assuming higher order relationships between the dependent and independent variables. For example assume the following linear model:

$$Y = a_1X_1 + a_2X_2 + e \qquad [9.1]$$

where Y is the dependent variable, X_1 and X_2 are independent variables, a_1 and a_2 are regression coefficients, and e is the residual. If the relationship between the dependent and independent variables is not linear, than the explanatory power of the model can be improved by inclusion of higher order terms as follows:

$$Y = a_1X_1 + a_2X_2 + a_3X_1^2 + a_4X_2^2 + a_5X_1X_2 + e \qquad [9.2]$$

Where the a's are regression coefficients and the other terms are as described previously. All possible second order regressions have been included in equation

[9.2]. If the dependent variable is the economic criterion (profit or economic efficiency) and the X's are the trait values, then linear selection index will be appropriate for the model of equation [9.1], but not for the model of equation [9.2].

In Chapter 6 we demonstrated that the economic values of different traits will be functions of the trait values. For example, in Table 6.4, the economic value of egg weight, computed on the basis of profit per hen, is a function of eggs per hen times egg weight. This is comparable to the fifth term in equation [9.2]. Although no examples are presented in Chapter 6, in which the economic value is a quadratic function of the trait value, many other functions can be approximated by a quadratic equation.

Wilton, Evans, and Van Vleck (1968) presented the following quadratic model of the aggregate genotype:

$$H_q = a'(\mu + y) + (\mu + y)'A(\mu + y) \tag{9.3}$$

Where H_q is the aggregate quadratic genotype of the individual; a is an $m \times 1$ vector of linear economic weights, as previously; μ is $m \times 1$ vector of trait means; y is a $m \times 1$ vector of breeding values, that is the genetic deviations from the trait means for a given animal; and A is the $m \times m$ matrix of quadratic economic weights. A is computed as follows:

$$A = \begin{bmatrix} a_{11} & 1/2a_{12} & \cdots & 1/2a_{1m} \\ 1/2a_{21} & a_{22} & & \vdots \\ \vdots & & & \vdots \\ 1/2a_{1m} & \cdots & & a_{mm} \end{bmatrix} \tag{9.4}$$

Where a_{ii} is the quadratic economic coefficient for the ith trait, and a_{ij} is the economic coefficient for the combination of the ith and j^{th} traits. Since these elements appear in the matrix both above and below the diagonal, they are multiplied by one half in the matrix. Thus the aggregate quadratic genotype differs from the linear aggregate genotype, both in the inclusion of the second term, and in the inclusion of the trait means. (The importance of the latter consideration will become clear shortly.)

9.3   Derivation of optimum linear and quadratic selection indices for quadratic models of the aggregate genotype

Using the quadratic aggregate genotype, we will first derive the optimum *linear* index, and then derive the optimum *quadratic* index. The linear index, I_1, was defined in Section 3.3 as follows:

$$I_1 = \mathbf{b'x} \qquad [9.5]$$

where $\mathbf{b}$ is an $n \times 1$ vector of index coefficients, $\mathbf{x}$ is an $n \times 1$ vector of trait values, expressed as deviations from the mean, and n is the number of traits included in the index. The optimum linear selection index will be defined as the index that minimizes the expectation of the squared difference between the aggregate genotype (which is in this case a quadratic function of the trait values), and the index. The deviation of the aggregate genotype from its expectation, $E(H_q)$ is computed as follows:

$$H_q - E(H_q) = (\mu + y)'\mathbf{a} + (\mu + y)'A(\mu + y) - [\mu'\mathbf{a} + \mu'A\mu + tr(AG)]$$

$$= y'\mathbf{a} + 2y'A\mu + y'Ay - tr(AG) \qquad [9.6]$$

Where "tr" denotes the trace of a matrix, and G is the genetic variance matrix among the traits. The expectation of the squared difference between the index and the aggregate genotype is then computed as follows:

$$E\{[I_1 - E(I_1)] - [H_q - E(H_q)]\}^2 = E[\mathbf{b'x} - y'\mathbf{a} - 2y'A\mu - y'Ay + tr(AG)]^2$$

$$= E[\mathbf{b'xx'b} - 2\mathbf{b'xy'a} - 4\mathbf{b'xy}A\mu - 2\mathbf{b'xy}Ay + 2\mathbf{b'x}\ tr(AG)]$$

$$+ \text{ terms not including } \mathbf{b}$$

$$= \mathbf{b'Pb} - 2'\mathbf{bCa} - 4\mathbf{b'C}A\mu + E[\text{terms not involving } \mathbf{b}] \qquad [9.7]$$

Where P is the phenotypic variance matrix, and C is the covariance matrix between $\mathbf{x}$ and y. (The expectation of $2\mathbf{b'xy}Ay + 2\mathbf{b'x}\ tr(AG)$ will be equal to zero.) Differentiating with respect to $\mathbf{b}$ and equating to zero gives:

$$2Pb = 2Ca + 4CA\mu \qquad [9.8]$$

$$\mathbf{b} = P^{-1}C[\mathbf{a} + 2A\mu] \qquad [9.9]$$

Thus the optimum linear index for the quadratic aggregate genotype is:

$$I_1 = \{P^{-1}C[\mathbf{a} + 2A\mu]\}'\mathbf{x} \qquad [9.10]$$

Note that the difference between this index and the standard linear index is the inclusion of the term $2A\mu$. Thus the optimum linear index is a function of the trait means for the given population. This explains the importance of inclusion of the means in the computation of the quadratic aggregate genotype in equation [9.3]. Contrary to the standard linear index, the index coefficients for I_l will therefore be functions of the trait means. (This question will be considered in more detail in the next chapter.)

The optimum quadratic index, I_q, is defined as follows:

$$I_q = b'x + x'Bx \qquad [9.11]$$

Where **B** is an n × n matrix of the form:

$$B = \begin{bmatrix} b_{11} & 1/2b_{12} & \dots & 1/2b_{1m} \\ 1/2b_{21} & b_{22} & & \vdots \\ \vdots & & & \vdots \\ 1/2b_{1m} & \dots & & b_{mm} \end{bmatrix} \qquad [9.12]$$

Where b_{ii} is the quadratic index coefficient for the ith trait, and b_{ij} is the economic coefficient for the combination of the ith and j^{th} traits. Similar to the matrix **A**, these elements appear in the matrix both above and below the diagonal, and are therefore multiplied by one half in the matrix.

The same method used above to derive I_l will now be used to derive I_q, i.e., minimization of the expectation of the squared deviation between the aggregate genotype and the index, both expressed as deviations from their expectations.

$$E\{[I_q - E(I_q)] - [H_q - E(H_q)]\}^2 =$$

$$= E[b'x + x'Bx - tr(BP) - y'a - 2y'A\mu - y'Ay + tr(AG)]^2 \qquad [9.13]$$

After several simplifications, and elimination of all terms which either have expectation of zero, or do not involve either **b** or **B**, this expectation can be rewritten as follows:

$$E\{[I_q - E(I_q)] - [H_q - E(H_q)]\}^2 =$$

$$= b'Pb - 2b'Ca - 4b'CA\mu + 2tr(BPBP) - 4tr(BCAC') \qquad [9.14]$$

Differentiating with respect to b and B and equating to zero gives:

$$\delta b/\delta E(.)^2 = 2Pb - 2C(a + 2A\mu) = 0 \qquad [9.15]$$

$$\delta B/\delta E(.)^2 = 4PBP - 4CAC' = 0 \qquad [9.16]$$

where $E(.)^2$ represents the expectation of the squared deviation between the aggregate genotype and the index. Solving for b and B we have:

$$b = P^{-1}C(a + 2A\mu) \qquad [9.17]$$

$$B = P^{-1}CAC'P^{-1} \qquad [9.18]$$

I_q is then computed as follows:

$$I_q = x'P^{-1}C(a + 2A\mu) + x'P^{-1}CAC'P^{-1}x \qquad [9.19]$$

I_q is the maximum likelihood estimate of H_q, conditional on x (Wilton, Evans, and Van Vleck, 1968).

9.4 Comparison of linear and quadratic indices for quadratic aggregate genotype functions

Although equation [9.19] is different from equation [9.10], it is of interest to determine what advantage will be gained, if any, by selection based on I_q, as opposed to I_l. In Section 3.5 we showed that the response to selection for the optimum linear index will be proportional to the standard deviation of the index. This will also be the case for the quadratic index. Thus the relative efficiency of I_l to I_q can be computed as the ratio of the standard deviations of the two indices. The variances of the two indices are computed as follows:

$$\sigma_{I_l}^2 = E[I_l - E(I_l)]^2 = b'Pb \qquad [9.20]$$

$$\sigma_{I_q}^2 = E[I_q - E(I_q)]^2 = b'Pb + 2\text{tr}(BPBP) \qquad [9.21]$$

Thus the relative selection efficiency (RSE) of the I_l is computed as follows:

$$RSE = \{b'Pb/[b'Pb + 2\ \text{tr}(BPBP)]\}^{1/2} = [1 + 2\ \text{tr}(BP)^2/b'Pb]^{-1/2} \qquad [9.22]$$

Application of quadratic indices will now be demonstrated using the example of Wilton, Evans, and Van Vleck (1968) for Angus beef cattle. They assume two traits with economic value: weaning weight and type score, and the following aggregate genotype function:

$$H_q = \$0.11(\mu_1 + y_1) + (\mu_1 + y_1)\$0.0049(\mu_2 + y_2) \qquad [9.23]$$

where μ_1 and y_1 refer to lbs weaning weight, and μ_2 and y_2 refer to points type score. In this example, the economic value for type score is a function of weaning weight, and a linear selection index is not appropriate. The means, phenotypic variances, and genetic variances of weaning weight and type score are respectively 419 lbs, and 13.35 points, 2649 lbs^2, and 1.75 points^2; and 1452 lbs^2 and 1.12 points^2. The phenotypic and genetic covariances between weaning weight and type score are 18.49 and 7.2 lbs points. Substituting these values into equation [9.9] the elements of **b** are computed as \$0.095/lb and \$1.0354/point, and the optimum linear index is:

$$I_1 = \$0.095x_1 + 1.0354x_2 \qquad [9.24]$$

where x_1 and x_2 are the individual's weaning weight and type score, respectively. To compute the optimum quadratic index, it is necessary to compute **B** from equation [9.18]. Substituting the values given above into equations [9.18] and [9.19], and simplifying gives the follow result:

$$I_q = \$0.0950x_1 + \$1.0354x_2 - 0.0000052x_1^2 - 0.0058551x_2^2$$

$$+ 0.0018302x_1x_2 \qquad [9.25]$$

Using equation [9.22] the relative efficiency of the linear index to the quadratic index is computed as 0.9997. This result is not surprising in view of the small index coefficients for the quadratic terms in equation [9.25]. It should be noted that, in general, the relative efficiency of I_1 will be very close to I_q for most practical situations. Wilton, Evans and Van Vleck (1968) noted that the quadratic index has an advantage in utility "because in the equivalent maximum likelihood development the estimates of genetic value for each trait can be made, and the economic values and the means can be changed, depending on the situation." This conclusion will be considered in more detail in the following chapter.

9.5 Restricted selection indices, theory

Because of the difficulties involved in accurately estimating the relative economic values of different traits, Kempthorne and Nordskog (1959) suggested the application of restricted selection indices as an alternative to linear selection index. The concept behind restricted selection indices can be stated as follows. Although we may not know the current economic values of unit changes in all traits, we do know that over the long-term certain changes would be very undesirable. Thus, the selection index is *restricted* within the "acceptable" parameter space of genetic changes. For example, a number of studies have been published on the genetic relationship between milk production and cow fertility (e. g. Hansen Freeman, and Berger, 1983; Hermas, Young, and Rust, 1987; Weller, 1989). Most of these studies have found that the genetic correlation between these traits is negative. Thus selection for production should produce a genetic reduction in fertility. Clearly this result would be undesirable. The linear selection approach would be to compute the optimum index based on the relative economic values of the two traits. However in doing this we would have to overcome all of the difficulties described above. In the restricted selection approach, we could assume *a priori* that any reduction in fertility is unacceptable. Under the restriction of zero genetic change in fertility, we then compute the index that maximizes genetic increase in milk production.

Most applications of restricted selection index have dealt with restrictions of this type, i.e. the index that gives maximum change in some traits under the restriction of zero change in other traits. Two other types of restricted indices have been proposed: fixed proportional change among traits (Pesek and Baker, 1969; Yamada, Yokouchi, and Nishida, 1974; Brascamp, 1984; Essl, 1981), and fixed absolute change among traits (Harville, 1975). In either case it is possible to restrict only some of the traits, and include the objective of maximum genetic change for the unrestricted traits. As an example of the applicability of this type of restriction, we can consider fat and protein content of milk. The ratio of fat to protein production of different dairy strains and countries ranges from 1 to 1.6, and this ratio can be affected by breeding (Gibson, 1987, 1989). The optimum selection index for these two traits can be readily computed based on the net economic value of each component. However, application of the optimum linear index may result in a change in the ratio of fat to protein production. Clearly it is desirable that the ratio of production should approximate the demand for any given country. Thus an index that maximizes production of both components, but restricts the change in ratio may be desirable. To derive the formula for optimum restricted indices, we will first review the derivation of the optimum unrestricted linear index.

9.6 Derivation of the optimum unrestricted selection index

In Chapter 3 we derived the optimum selection index coefficients by minimizing the squared difference between **I** and **H**. Brascamp (1984) presents two additional methods to derive this index. They are: 1) to maximize the average expected breeding values of the selected individuals, and 2) to maximize the correlation between **I** and **H**. One of the major strengths of linear selection index is that the same solution for **b** is obtained by all three methods.

We will now consider the method of maximization of expected breeding value in some detail for unrestricted index, before proceeding to the derivation of the restricted index. As shown in equation [3.26], the response to selection on a linear index, ϕ_I, is computed as follows:

$$\phi_I = ir_{HI}\sigma_H = i\sigma_{HI}/\sigma_I \qquad [9.26]$$

Where σ_I is the standard deviation of the index. (The additional subscript "s" in equation [3.26] is now deleted.) The selection intensity, i, is a constant, and can therefore be assumed to equal unity. From equations [3.24] and [3.25] we have:

$$\sigma_I^2 = \mathbf{b'Pb} \qquad [9.27]$$

$$\sigma_{HI} = \mathbf{b'Ca} \qquad [9.28]$$

The optimum values for **b** can then be computed by differentiating ϕ_I with respect to **b**, and equating to zero as follows:

$$d\phi_I/d\mathbf{b} = \mathbf{Ca}/\sigma_I - \mathbf{Pb}\sigma_{HI}/(\sigma_I)^3 = 0 \qquad [9.29]$$

$$[\sigma_{HI}/(\sigma_I)^2]\mathbf{b} = \mathbf{P}^{-1}\mathbf{Ca} \qquad [9.30]$$

Since we are interested only in the *relative* values of the elements of **b**, **b** can be scaled so that $\sigma_{HI}/(\sigma_I)^2 = b_{HI} = 1$. In this case we derive the result of equation [3.17]: $\mathbf{b} = \mathbf{P}^{-1}\mathbf{Ga}$. Alternatively we can set: $\sigma_I^2 = \mathbf{b'Pb} = 1$. In this case the solution for **b** is found by maximizing $\mathbf{b'Ca} - \tau(\mathbf{b'Pb} - 1)$ where τ is a "Lagrange multiplier". **b'Ca** does not have a unique maximum. However, **b'Ca** $- \tau(\mathbf{b'Pb} - 1)$ will be maximum only when $\mathbf{b'Pb} = 1$. The values of **b** are found by differentiating with respect to **b**, and equating to zero, and then solving for the Lagrange multiplier. This of course gives the same result for **b** as the previous methods, but has the useful property that the correlated response of an individual trait due to selection on the index will reduce to $[Cov(\mathbf{g_i,p'})]\mathbf{b}$ where $Cov(\mathbf{g_i,p'})$ is the ith column of **G**. If $\mathbf{G} = \mathbf{C}$, that is if all traits included in the genetic variance matrix are also included in the index, then the equation for the correlated responses for the individual traits given in [3.31] reduces to $\phi = \mathbf{b'G}$ per unit selection intensity.

9.7 Derivation of optimum restricted selection indices

We will divide the traits included in the index into two groups, $\mathbf{g_o}$ and $\mathbf{g_1}$, and assume that all traits included in the aggregate genotype are included in the index so that $\mathbf{C} = \mathbf{G}$. Consider the general case in which maximum gain is desired for $\mathbf{g_o}$ traits, while the relative genetic gain for the remaining $\mathbf{g_1}$ traits is restricted. We will partition **G** as follows: $\mathbf{G} = (\mathbf{G_o}:\mathbf{G_1})$, and assume that $\sigma_I = 1$. The correlated response of $\mathbf{g_o}$ will then be $\mathbf{G_o'b}$, and the correlated response of $\mathbf{g_1}$ will be G_1'b. The restriction can then be defined as follows:

$$\phi = \alpha \ \delta \qquad\qquad\qquad [9.31]$$

$$\phi = G_1'b \qquad\qquad\qquad [9.32]$$

where ϕ is a vector of absolute changes for the traits in g_1, α is a proportionality constant, and δ is vector of the proportional changes for the traits in g_1.

The optimum restricted index can then be computed in terms of ϕ (fixed absolute trait changes) or δ (fixed relative trait changes). We will first solve in terms of ϕ and then substitute to solve for δ. Defining a_0 as the economic value of the traits in g_0, the optimum restricted index is then computed by maximizing b'$G_0 a_0$, under the conditions that G_1'b = ϕ, and σ_I = 1. In this case, the function to be differentiated is: b'$G_0 a_0$ − Γ'(G_1'b − ϕ) − τ(b'Pb − 1). Where Γ is a vector of Lagrange multipliers and τ is a scalar Lagrange multiplier. Differentiating and equating to zero gives:

$$G_0 a_0 \ - \ G_1\Gamma \ - \ 2\tau Pb = 0 \qquad\qquad\qquad [9.33]$$

It is then necessary to solve for Γ and τ in terms of the other variables. The matrix algebra for this is quite complex and is given in Brascamp (1984). We will only present the final solution for b:

$$b = \frac{(1 \ - \ \phi'(G_1'P^{-1}G_1)^{-1}\phi)^{1/2}}{(a_0'G_0 RG_0 a_0)^{1/2}} \ RG_0 a_0 + P^{-1}G_1(G_1'P^{-1}G_1)^{-1}\phi \qquad [9.34]$$

where:

$$R = P^{-1}(I \ - \ G_1(G_1'P^{-1}G_1)^{-1}G_1'P^{-1} \qquad\qquad\qquad [9.35]$$

A solution will exist only if $1 \ - \ \phi'(G_1'P^{-1}G_1)^{-1}\phi > 0$. If this term is equal to zero, then the change in g_1 will be maximum, and the change in g_0 will be zero.

We will now solve for ϕ in terms of δ, with the aid of the following equation:

$$\phi_{max} = \alpha_{max} \ \delta \qquad\qquad\qquad [9.36]$$

where ϕ_{max} is the maximum possible value for ϕ, and α_{max} is a proportionality constant. Substituting into equation [9.34] and simplifying gives:

$$b = \frac{(1 \ - \ \alpha^2/\alpha^2_{max})^{1/2}}{(a_0'G_0 RG_0 a_0)^{1/2}} \ RG_0 a_0 + \alpha P^{-1}G_1(G_1'P^{-1}G_1)^{-1}\delta \qquad [9.37]$$

This equation no longer contains ϕ. The optimum restricted index is now computed as a function of δ, which is assumed to be known, α_{max}, which can be computed from equation [9.36], and α. Although α is not known, it can be chosen so that $0 < \alpha \leq \alpha_{max}$.

We will now develop the equations for two specific types of restricted indices; indices with the change in some traits restricted to zero, and selection indices with all traits restricted to specified changes. For restriction of some traits to zero, $\alpha = 0$ and $\delta = \phi = 0$. In this case, equation [9.34] reduces to:

$$\mathbf{b} = (\mathbf{a_o'G_oRG_oa_o})^{-1/2}\mathbf{RG_oa_o} \qquad [9.38]$$

Since $(\mathbf{a_o'G_oRG_oa_o})^{-1/2}$ is a scalar, a new vector, ß can be defined proportional to $\mathbf{b}$ and equal to:

$$\text{ß} = \mathbf{RG_oa_o} \qquad [9.39]$$

Cunningham, Moen, and Gjedrem (1970) solved for ß using the following normal equations:

$$\begin{bmatrix} \mathbf{P} & \mathbf{G_1} \\ \mathbf{G_1'} & 0 \end{bmatrix} \begin{bmatrix} \text{ß} \\ \Gamma \end{bmatrix} = \begin{bmatrix} \mathbf{G_oa_o} \\ 0 \end{bmatrix} \qquad [9.40]$$

Where Γ is a vector of Lagrange multipliers, that solve the equations $\mathbf{G_1'}\text{ß} = 0$. Cunningham, Moen, and Gjedrem (1970) also noted that there exists a vector of economic weights, $\mathbf{a_1}$, that would have resulted in zero changes for the traits in $\mathbf{g_1}$ with the optimum unrestricted index. The solution for $\mathbf{a_1}$ is $-\Gamma$. Thus as a check on the validity of this index, these "pseudo" economic weights can be computed, and their values compared to the best available estimates for $\mathbf{a_1}$. Restricted indices of this type were developed for dairy cattle by Niebel and Van Vleck (1982, 1983), in which the genetic change in calving difficulty was restricted to zero.

The optimum selection index for desired changes is merely the specific case where all traits are included in $\mathbf{g_1}$. In this case, equation [9.34] reduces to:

$$\mathbf{b} = \mathbf{P^{-1}G_1(G_1'P^{-1}G_1)^{-1}}\phi = \alpha_{max}\mathbf{P^{-1}G_1(G_1'P^{-1}G_1)^{-1}}\delta \qquad [9.41]$$

Since all traits are restricted, it is necessary only to compute the relative index weights of the traits. Therefore, α_{max}, which is now a proportionality constant, can be dropped. For selection on individual performance, and the same traits in I and H, that is $\mathbf{C_1} = \mathbf{G_1}$, equation [9.41] then simplifies as follows (Pesek and Baker, 1969):

$$b = G_1^{-1}\delta = C_1^{-1}\delta \qquad\qquad [9.42]$$

We can compute, as in the previous case, the values for a_1 that would give the same unrestricted selection index. From equation [9.32] we have $\phi = G_1'b$, and for linear selection index $b = P^{-1}G_1a_1$. Substituting for b gives:

$$\phi = G_1'P^{-1}G_1a_1 \qquad\qquad [9.43]$$

$$a_1 = (G_1'P^{-1}G_1)^{-1}\phi \qquad\qquad [9.44]$$

Again these "pseudo" economic values can be checked against the best guess available as to the true economic values. Finally we should note that although restricted indices are linear, they will be less efficient than the optimum linear index, assuming the economic weights are known, and that they are linear functions of the trait values.

We will illustrate restricted indices with the following example from milk production. As stated previously, the genetic correlations between milk production and fat and protein percent are negative. Furthermore although in many countries, protein is more valuable than fat, the standard deviation and heritability of protein are lower. Assume that the breeding objective is to maximize protein and fat production, under the constraints that genetic gain in both traits will be equal, and that the mean percent of protein and fat will not change from the current value of 3.3%. That is:

$$\phi_m = (\phi_p + \phi_f)/0.066 \qquad\qquad [9.45]$$

where ϕ_m, ϕ_p and ϕ_f are the genetic gains in milk, fat and protein, respectively. We will further assume that the selection index will consist only of these three traits, and that their genetic and phenotypic variance matrices are known to be as follows:

$$
G = \begin{bmatrix} 527{,}148 & 6756 & 9973 \\ 6756 & 502 & 249 \\ 9973 & 249 & 306 \end{bmatrix} \qquad\qquad [9.46]
$$

$$
P = \begin{bmatrix} 1{,}469{,}190 & 32{,}883 & 37{,}943 \\ 32{,}883 & 1557 & 1023 \\ 37{,}943 & 1023 & 1220 \end{bmatrix} \qquad\qquad [9.47]
$$

The elements of δ will then be 1/0.033, 1, 1. Assuming selection on individual performance, equation [9.42] can be used to solve for b. G^{-1} will be:

$$\mathbf{G}^{-1} = \begin{bmatrix} 5.10 & 23.17 & -185.19 \\ 23.17 & 3445.42 & -3558.2 \\ -185.19 & -3558.2 & 12199.4 \end{bmatrix} (10^{-6}) \qquad [9.48]$$

The elements of **b** are then computed as $\mathbf{G}^{-1}\boldsymbol{\delta}$, and are equal to $(-0.0071, 0.59,$ and $3.03) \times 10^{-3}$. The vector **a** can then be computed from equation [9.44], and its values are $(-0.027, 1.56, 12.7) \times 10^{-3}$. Scaling these values so that the economic value from a kg of milk with mean protein and fat content should equal unity, gives: $\mathbf{a'} = (-0.07, 5.7, 29.2)$.

A first glance, this result is somewhat surprising. Even though fat and protein production have similar heritabilities and means, and the goal was equal increase in the two components, the "pseudo" economic value of protein to achieve this objective is five−fold the economic value of fat. Furthermore, even though all three traits are positively correlated, and the objective is to increase milk production, the economic value of milk is negative. These results can be understood in the light of the genetic variance and covariances. The genetic variance for fat is greater than for protein and the genetic correlation between these traits is 0.54. Likewise the genetic correlation between protein and milk production is 0.78. Thus direct selection on protein will increase fat production nearly as much as protein, and will increase milk production slightly more than the desired objective. Thus the optimum index will consist chiefly of selection for protein, with a small positive coefficient for fat, and a small negative coefficient for milk.

We can compare these results to the linear selection index obtained under the assumption of zero economic weight for milk, and equal economic weight for fat and protein. In this case, $\mathbf{a'} = [0 \quad 1 \quad 1]$, and $\mathbf{b'} = [-0.0033 \quad 0.40 \quad 0.22]$. The vector of correlated responses with i = 2 will be $\boldsymbol{\phi'} = [340.8 \quad 24.4 \quad 14.14]$. δ times 14.14 gives: [428 14.14 14.14]. Thus the vectors of correlated responses were much more similar than the vectors of economic weights. This result will be generally true, and will be considered in more detail in Chapter 10.

9.8 Selection indices by graphic methods, basic concepts

The graphic method was developed by Moav and Hill (1966) for the case of two traits, and both linear and nonlinear profit functions. Although graphs were used extensively by Moav and Hill (1966) to illustrate the method, it can also be phrased algebraically using principles of analytical geometry. We will first assume that profit is a linear function of two uncorrelated traits, x and y. After the explanation for this case, the method will then be generalized step-by-step.

The index coefficients for any possible index, I, are denoted b_x and b_y, so that:

$$I = b_x x + b_y y \qquad\qquad [9.49]$$

The variance for a selection index was given as **b'Pb** in equation [3.24]. For two uncorrelated traits, this reduces to:

$$(\sigma_I)^2 = (b_x \sigma_x)^2 + (b_y \sigma_y)^2 \qquad\qquad [9.50]$$

The response to selection of either trait, due to selection on the index can be computed from equation [3.16], for the correlated response. That is:

$$\phi_x = i h_I r_g \sigma_{AX} \qquad\qquad [9.51]$$

where ϕ_x is the response of x, h_I is the accuracy of the index, r_g is the genetic correlation between the index and x, and σ_{AX} is the additive genetic variance of x. $h_I = \sigma_{AI}/\sigma_I$, where σ_{AI} is the additive genetic standard deviation of the index. The genetic covariance between x and the index will be $b_x(\sigma_{AX})^2 = b_x(h_x\sigma_x)^2$. Thus the correlated responses of x and y can be computed as follows:

$$\phi_x = i b_x (h_x \sigma_x)^2 / \sigma_I \qquad\qquad [9.52]$$

$$\phi_y = i b_y (h_y \sigma_y)^2 / \sigma_I \qquad\qquad [9.53]$$

Equation [9.50] can be then be rewritten as follows:

$$(\sigma_I)^2 = \left[\frac{i b_x h_x^2 \sigma_x^2}{\sigma_I} \right]^2 \frac{\sigma_I^2}{(i h_x^2 \sigma_x)^2} + \left[\frac{i b_y h_y^2 \sigma_y^2}{\sigma_I} \right]^2 \frac{\sigma_I^2}{(i h_y^2 \sigma_y)^2} \qquad [9.54]$$

Substituting equations [9.52] and [9.53] into equation [9.54] and dividing both sides by σ_I^2 gives:

$$1 = \frac{(\phi_x)^2}{(i h_x^2 \sigma_x)^2} + \frac{(\phi_y)^2}{(i h_y^2 \sigma_y)^2} \qquad\qquad [9.55]$$

All terms in equation [9.55] are constants, except ϕ_x, ϕ_y, and i. Thus for a given selection intensity, equation [9.55] describes an ellipse in the variables ϕ_x

and ϕ_y, and is termed the "response ellipse" (Moav and Hill, 1966).

In Section 6.6 we demonstrated how profit can be represented graphically by a profit map. A simple profit map has been plotted in Figure 9.1, in which the change in profit, ϕ_P, is a linear function of two traits. That is:

$$\phi_P = a_x \phi_x + a_y \phi_y \qquad\qquad [9.56]$$

$$\phi_y = \phi_P/a_y - \phi_x a_x/a_y \qquad\qquad [9.57]$$

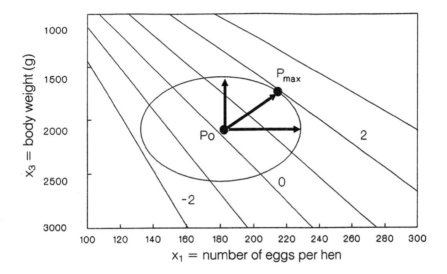

Figure 9.1. Profit map and response ellipse for laying hens. Vertical and horizontal arrows are response to direct selection on x_3 and x_1, respectively. P_{max} is the point of maximum gain in profit due to selection.

where a_x and a_y are the economic values for x and y. The slope of the profit contours will be $-a_x/a_y$. The response ellipse is also illustrated in this figure. The axes of this ellipse will be equal to $ih_x^2\sigma_x$ and $ih_y^2\sigma_y$, and will be parallel to the x and y axes. As illustrated in Figure 9.1, a number of profit contours cut the response ellipse. All values for ϕ_P that meet the response ellipse can be obtained by a given combination of b_x and b_y. Note that the profit contours are tangent to the response ellipse at only two points. These will be the points of maximum positive and negative ϕ_P obtainable by selection on I. At these points the direction of response will be perpendicular to the response contours. Of course, only the point of maximum positive profit is of interest. This point is found by equating the slope of the profit contour, $-a_x/a_y$, to the slope of the ellipse for a given value of x and y. The slope of the ellipse is computed as: $\delta d/\delta x = -\phi_x h_y^4 \sigma_y^2/(\phi_y h_x^4 \sigma_x^2)$. Thus, the values for maximum genetic gain are computed as follows:

$$\frac{a_x}{a_y} = \frac{\phi_x h_y^4 \sigma_y^2}{\phi_y h_x^4 \sigma_x^2} \qquad [9.58]$$

Substituting for ϕ_x and ϕ_y in equations [9.52] and [9.53] and simplifying gives:

$$\frac{b_y}{b_x} = \frac{a_y h_y^2}{a_x h_x^2} \qquad [9.59]$$

The selection index coefficients for an index of two uncorrelated traits will be $a_x h_x^2$ and $a_y h_y^2$. Thus, for the case of a linear profit function, the graphic method gives the same result as standard selection methodology. In summation the graphic method computes the optimum selection index by equating the tangent of the response ellipse to the profit function, and solving for the elements of **b** in terms of the other parameters.

Further calculations are simplified somewhat if the trait units are converted into standardized response units as follows:

$$\phi_x^* = \phi_x/(h_x^2 \sigma_x), \ \phi_y^* = \phi_y/(h_y^2 \sigma_y) \qquad [9.60]$$

where ϕ_x^* and ϕ_y^* are the responses of x and y to selection in standardized response units. Equation [9.55] can then be rewritten as follows:

$$(\phi_x^*)^2 + (\phi_y^*)^2 = i^2 \qquad [9.61]$$

Equation [9.61] defines a circle that Moav and Hill (1966) termed the response circle. The economic values of ϕ_x^* and ϕ_y^* will be defined as follows:

$$a_x^* \phi_x^* = a_x \phi_x, \ a_y^* \phi_y^* = a_y \phi_y \qquad [9.62]$$

Then substituting from equations [9.60] gives:

$$a_x^* = a_x h_x^2 \sigma_x, \ a_y^* = a_y h_y^2 \sigma_y \qquad [9.63]$$

The index for maximum profit is again computed by equating the tangent of the profit contours and the response circle, both in standardized units:

$$\phi_y^*/\phi_x^* = a_y^*/a_x^* \qquad [9.64]$$

Substituting equations [9.63] into [9.64] gives:

$$\phi_y^*/\phi_x^* = a_y h_y^2 \sigma_y / a_x h_x^2 \sigma_x \tag{9.65}$$

Finally substituting equations [9.52] and [9.53] into equation [9.65] gives:

$$b_y/b_x = \phi_y^* \sigma_x / \phi_x^* \sigma_y \tag{9.66}$$

which is, of course, the selection index result for the transformed traits.

9.9 Selection indices by graphic methods for two correlated traits and nonlinear profit functions

Moav and Hill (1966) extended the graphic method to cover the case of two correlated traits and a nonlinear profit function. Again we will assume that all traits in the aggregate genotype are included in the selection index, so that **G** = **C**. (This restriction will be removed below.) The vector of correlated responses, ϕ, is then computed as in equation [3.31]:

$$\phi = iGb/\sigma_I \tag{9.67}$$

From equation [3.24] we have:

$$\sigma_I^2 = b'Pb = b'GG^{-1}PG^{-1}Gb \tag{9.68}$$

Substituting equation [9.67] into equation [9.68] gives the following result:

$$\phi'G^{-1}PG^{-1}\phi/i^2 = 1 \tag{9.69}$$

For two traits, this function defined a response ellipse that differs from the ellipse of equation [9.55] in that in addition to the two terms that include ϕ_x^2 and ϕ_y^2, there will be a third term that includes $\phi_x\phi_y$. Thus the axes of this ellipse will not be parallel to the coordinate axes. For more than two traits, equation [9.69] defines an n-dimensional ellipse, where n is the number of traits in the index. The optimum index can be derived as in the previous case by equating the slope of the response ellipse and the profit function.

The response ellipse of equation [9.69] can be converted into a circle by the following transformation: Define a matrix **J** so that **J'J** = **P**. Then equation [9.69] can be rewritten as follows:

$$\phi'G^{-1}J'JG^{-1}\phi = i^2 = \phi^{*'}\phi^* \tag{9.70}$$

where $\phi^* = JG^{-1}\phi$. Equation [9.70] now defines a response circle for the case of two traits, or a multidimensional response circle for the multitrait case. Parallel to the previous case we will define $a^* = J'^{-1}Ga$. For a matrix of two traits:

$$J = \begin{bmatrix} \sigma_x \cos\theta & \sigma_y \sin\theta \\ \sigma_x \sin\theta & \sigma_y \cos\theta \end{bmatrix} \qquad [9.71]$$

Where $r_{xy} = \sin 2\theta$. To derive this matrix, we remind the reader that:

$$\sin 2\theta = 2 \sin\theta \cos\theta \qquad [9.72]$$

$$\sin^2\theta = 1/2 - (1/2)\cos 2\theta \qquad [9.73]$$

$$\cos^2\theta = 1/2 + (1/2)\cos 2\theta \qquad [9.74]$$

We will now consider the case of a nonlinear profit function, first with uncorrelated traits, and then with correlated traits. In equation [6.21] and [7.9] we presented the following profit function:

$$P_1 = K_1 - K_2x_2 - K_3/x_1 \qquad [9.75]$$

where P_1 is profit per pig marketed, x_1 is number of pigs weaned per sow per year, x_2 is age to a fixed market weight, K_1 is income less costs independent of x_1 and x_2, K_2 is costs dependent on x_2, and K_3 is fixed costs (feed and non-feed) per sow. In this equation profit is an inverse function of x_1. Thus none of the methods presented previously can be used to compute the optimum selection index. Setting $x_1 = x$ and $x_2 = y$, the partial differentials of P_1 are K_3/x^2 and $-K_2$ for x and y, respectively. Assuming that these traits are uncorrelated, equation [9.59] can be used to compute the optimum index, with a_y/a_x replaced by $-x^2K_2/K_3$, as follows:

$$\frac{b_y}{b_x} = \frac{-x^2K_2h_y^2}{K_3h_x^2} \qquad [9.76]$$

Thus the optimum index for a nonlinear profit function is the same as the optimum linear index with the linear economic values replaced by the differentials of the profit function. For two correlated traits, the optimum index can be also be derived by equating tangents for the response circle and the economic values for the standardized trait units, that is $a^* = (J')^{-1}Ga$. After re-arranging, we derive:

$$\frac{b_y}{b_x} = \frac{\sigma_x[1 + (1 - r^2)^{1/2}]\phi_y^* - r\phi_x^*}{\sigma_y[1 + (1 - r^2)^{1/2}]\phi_x^* - r\phi_y^*} \tag{9.77}$$

which reduces to equation [9.66] for r = 0.

9.10 Selection indices by graphic methods for more than two traits

Pasternak and Weller (1993) extended the method of Moav and Hill (1966) to cover any number of correlated traits, and also situations where some traits in the profit function are not included in the selection index, and *vice versa*. We will first consider the case in which all traits in the profit function are included in the selection index. Phrased in terms of nonlinear programming the problem can be stated as follows:

$$\text{Maximize: } f(\mathbf{X} + \phi) \tag{9.78}$$

$$\text{Subject to: } \phi'\mathbf{G}^{-1}\mathbf{P}\mathbf{G}^{-1}\phi \le i^2 \tag{9.79}$$

Where $f(\mathbf{X} + \phi)$ is the profit function of the vector of economic traits after selection, $\mathbf{X}$ is the vector of trait means prior to selection, and the other terms are as defined above. It is assumed that $f(\mathbf{X} + \phi)$ is twice continuously differentiable and concave with respect to ϕ, and that $\mathbf{P}$ is real, symmetric and positive definite. Under these assumptions, the matrix of second differentials of $f(\mathbf{X} + \phi)$ is negative definite (Graybill, 1969), and $\mathbf{G}^{-1}\mathbf{P}\mathbf{G}^{-1}$ will be positive definite (Schmidt, 1982), and convex in ϕ (Jacoby, Kowalik, and Pizzo, 1972). (This means that the matrix of second differentials for i^2 with respect to ϕ is positive definitive.) Since inequality [9.79] is a less than or equal constraint, the "feasible" parameter space will be convex (Peressini, Sullivan, and Uhl, 1988). Hence, the Kuhn-Tucker conditions are necessary and sufficient conditions for maximization of the objective function (Nemhauser, Rinnooy Kan, and Todd, 1989). Applying the Kuhn-Tucker conditions to the objective and constraint function yields the following equations:

$$\{\delta[f(\mathbf{X} + \phi^*)]/\delta\phi\}' = 2\Gamma\phi^{*'}\mathbf{G}^{-1}\mathbf{P}\mathbf{G}^{-1} \tag{9.80}$$

Where $\delta[f(\mathbf{X} + \phi^*)]/\delta\phi$ is the vector of partial derivatives of $f(\mathbf{X} + \phi^*)$ with respect to ϕ; ϕ^* are the values for ϕ that maximize $f(\mathbf{X} + \phi)$, subject to inequality [9.79]; and Γ is a scalar, and equal to: $\delta[f(\mathbf{X} + \phi^*)]/\delta i^2$. That is, the

objective function will be maximized over the feasible parameter space if the
vector of the partial derivatives of the objective function with respect to ϕ is
equal to the derivative of the profit function with respect to i^2, times the vector
of the partial derivatives of the constraint function, also with respect to ϕ.
Solving equations [9.80] for ϕ^* gives:

$$\phi^* = (1/2\Gamma)GP^{-1}G\{\delta[f(X + \phi^*)]/\delta\phi\} \qquad [9.81]$$

All terms on the right-hand side of equations [9.81] are known, except for ϕ^* and
Γ.   The constraint given in inequality [9.79] will be used to solve for Γ.
Goddard (1983) considered two situations with respect to the location of the
maximum for $f(X + \phi)$ on the multitrait parameter space.   If ϕ^* is on the
multidimensional response ellipse surface, then under the constraint given above,
profit will be maximized by a linear selection index and maximum selection
intensity.   In this case equation [9.69] can be modified to solve for ϕ^* as follows:

$$\phi^{*'}G^{-1}PG^{-1}\phi^* = i^2 \qquad [9.82]$$

Although the Kuhn-Tucker conditions hold for inequality [9.79], it is easier to
derive solutions for an equality.   Using equation [9.81] to solve for ϕ^* in
equation [9.82], and re-arranging gives the following:

$$\Gamma = \frac{(\{\delta[f(X + \phi^*)]/\delta\phi\}'GP^{-1}G\{\delta[f(X + \phi^*)]/\delta\phi\})^{1/2}}{2i} \qquad [9.83]$$

Substituting this value for Γ in equations [9.81] and rearranging gives:

$$\phi^* = \frac{GP^{-1}G\{\delta[f(X + \phi^*)]/\delta\phi\}i}{(\{\delta[f(X + \phi^*)]/\delta\phi\}'GP^{-1}G\{\delta[f(X + \phi^*)]/\delta\phi\})^{1/2}} \qquad [9.84]$$

Equations [9.84] no longer contain Γ, but ϕ^* appears on both sides of the
equations. Thus these equations must be solved iteratively. Note also that ϕ^* is
a function of i. The selection intensity did not appear in the equations of Moav
and Hill (1966) for the optimum index for nonlinear profit functions and two
traits. However, even for two traits, the index coefficients were dependent on ϕ^*,
which is a function of i.

   ϕ^* can be solved by iteration on the following equations, derived from
equations [9.84]:

$$\phi^{k+1} = \frac{GP^{-1}G\{\delta[f(X + \phi^k)]/\delta\phi\}i}{(\{\delta[f(X + \phi^k)]/\delta\phi\}'GP^{-1}G\{\delta[f(X + \phi^k)]/\delta\phi\})^{1/2}} \qquad [9.85]$$

Where ϕ^k is the value of ϕ at the kth iteration, and ϕ^{k+1} is the value of ϕ after the ith iteration. Finally from equation [9.67], we have:

$$b = (\sigma_I/i)G^{-1}\phi^* \qquad [9.86]$$

Since only the relative values of the index coefficients are important (Moav and Hill, 1966), σ_I/i can be considered a proportionality constant, and $G^{-1}\phi^*$ can be scaled to any convenient value.

If $f(X + \phi)$ is a linear function of the trait values, then $\delta[f(X + \phi^*)]/\delta\phi = a$, where a is a vector of constants. Equations [9.84] then reduce to:

$$\phi^* = GP^{-1}Gai(a'GP^{-1}Ga)^{-1/2} \qquad [9.87]$$

Solving for ϕ^* from equation [9.86] gives:

$$iGb/\sigma_I = GP^{-1}Gai(a'GP^{-1}Ga)^{-1/2} \qquad [9.88]$$

$$b = P^{-1}Ga\sigma_I/(a'GP^{-1}Ga)^{1/2} \qquad [9.89]$$

Thus $b = P^{-1}Ga$ times a proportionality constant, $\sigma_I/(a'GP^{-1}Ga)^{1/2}$. Since $\sigma_I^2 = b'Pb$, this constant will be equal to unity for $b = P^{-1}Ga$. Thus for profit functions linear in the trait values, the general selection index reduces to the standard linear selection index equations.

9.11 Some traits in the profit function are not included in the selection index, and *vice versa*

The situation in which some traits included in the selection index have no direct economic value can be readily handled by setting $\delta[f(X + \phi^k)]/\delta\phi$ in equations [9.85] to zero for those traits. If some traits in the profit function are not included in the selection index, then inequality [9.79] must be modified as follows:

$$\phi'G_1^{-1}P_1G_1^{-1}\phi \leq i^2 \qquad [9.90]$$

Where G_1 and P_1 are the genetic and phenotypic variance matrices for those traits that are included in the selection index. The genetic gains for the optimum index can then be found by solving equations [9.84], modified as follows:

$$\phi_1^* = \frac{G_1 P_1^{-1} G_1 \{\delta[f(X + \phi_1^*)]/\delta\phi_1\} i}{(\{\delta[f(X + \phi_1^*)]/\delta\phi_1\}' G_1 P_1^{-1} G_1 \{\delta[f(X + \phi_1^*)]/\delta\phi_1\})^{1/2}} \qquad [9.91]$$

Where ϕ_1 is the vector of genetic gains for those traits included in the index. The genetic gains of the traits included in $f(X + \phi)$ that are not included in the index can be computed as functions of the genetic gains for the traits included in the index as follows: First the selection index for the traits in ϕ_1, b_1, can be computed from equation [9.86], modified as follows:

$$b_1 = (\sigma_I/i) G_1^{-1} \phi_1 \qquad [9.92]$$

From equation [9.86], the vector of genetic gains for the traits not included in the index, ϕ_2, can then be computed as:

$$\phi_2 = (i/\sigma_I) G_2 b_1 \qquad [9.93]$$

Where G_2 is a matrix consisting of the rows of G for those traits not included in the index, but with the columns corresponding to these traits deleted. Thus G_2 can be multiplied by b_1. Substituting for the value for b_1 from equation [9.92] in equation [9.93] gives: $\phi_2 = G_2 G_1^{-1} \phi_1$. Thus, $f(\phi) = f(\phi_1, \phi_2) = f(\phi_1, G_2 G_1^{-1} \phi_1) = f(\phi_1)$, it is not necessary to solve for σ_I, and ϕ_2 is a linear function of ϕ_1. The derivatives of the objective function can then be computed accordingly.

9.12 A numerical example with three correlated traits and a nonlinear profit function

The method will be illustrated using the example of Pasternak and Weller (1993) for the Israeli dairy industry. Profit as a function of milk (carrier), fat, and protein production was computed with the following equation:

$$F(X + \phi) =$$

$$\frac{(R_m - C_m)(X_m + \phi_m) + (R_f - C_f)(X_f + \phi_f) + (R_p - C_p)(X_p + \phi_p) - C_c}{X_f + \phi_f + X_p + \phi_p} \qquad [9.94]$$

where R_m, R_f, and R_p are returns for kg milk, fat, and protein; C_m, C_f, and C_p are costs proportional to the quantity produced of each component, X_m, X_f, and X_p are the mean production of kg milk, fat, and protein per cow per year, ϕ_m, ϕ_f, and ϕ_p are the expected genetic gains, and C_c are fixed costs per cow not dependent on production. Following Moav (1973), $F(X + \phi)$ is computed as profit per kg fat + protein, under the assumption that the future quota system will be based on this criterion. Thus, for the producer, the profit criterion of interest is profit per kg fat + protein. Inserting values appropriate for the Israeli dairy industry in 1991 (Weller and Ezra, 1991) gives:

$$F(X + \phi) =$$

$$= \frac{-0.18(9244 + \phi_m) + 4.2(290 + \phi_f) + 22.9(285 + \phi_p) - 5000}{290 + \phi_f + 285 + \phi_p} \qquad [9.95]$$

These values are based on an assumed price of 1 Israeli Shekel (IS) = $0.40 for a "standard" kg milk, with mean fat and protein concentration. C_c was set at 5000 IS so that profit for a cow with mean production after selection would be positive. This was done to avoid the situation considered by Brascamp, Smith, and Guy (1985) of zero net profit.

The partial differentials of equation [9.95] with respect to ϕ are:

$$\frac{\delta[F(X + \phi)]}{\delta\phi_m} = \frac{-0.18}{290 + \phi_f + 285 + \phi_p} \qquad [9.96]$$

$$\frac{\delta[F(X + \phi)]}{\delta\phi_f} = \frac{4.2[290 + \phi_f + 285 + \phi_p] - Q}{[290 + \phi_f + 285 + \phi_p]^2} \qquad [9.97]$$

$$\frac{\delta[F(X + \phi)]}{\delta\phi_p} = \frac{22.9[290 + \phi_f + 285 + \phi_p] - Q}{[290 + \phi_f + 285 + \phi_p]^2} \qquad [9.98]$$

Where:

$$Q = -0.18(9244 + \phi_m) + 4.2(290 + \phi_f) + 22.9(285 + \phi_p) - 5000 \qquad [9.99]$$

Note that equations [9.96], [9.97], and [9.98] are nonlinear with respect to ϕ_f and ϕ_p. $GP^{-1}G$ was computed as follows:

$$GP^{-1}G = \begin{bmatrix} 242,407 & 529 & 2760 \\ 529 & 179 & 57 \\ 2760 & 57 & 79 \end{bmatrix} \qquad [9.100]$$

These values were then used to iterate on equations [9.85] for two values of i: i = 1, and i = 4. ϕ^1 was set equal to a vector of zeros. Iteration was continued until all elements of $\varepsilon = |\phi^{k+1} - \phi^k|$ were less than the critical value, set at $\phi^k/1000$ for all elements of ϕ^k. Results are given in Table 9.1. Convergence was obtained after three iterations for i = 1, and after four iterations for i = 4. Equation [9.86] was then used to compute the optimum indices for each value of i. Index coefficients standardized to kg fat are given in Table 9.2. Also given are the expected genetic gains relative to the gain in fat. Although the relative genetic gains are very similar for both selection intensities, the index coefficients to achieve these gains are not. This will be true if the deviation from linearity is not extreme over the range of possible genetic progress, and corresponds to the results of previous studies, that relatively large changes in the index coefficients result in only minor changes in the efficiency of selection (Smith, 1983; Wilton, Evans, and Van Vleck, 1968).

Values for Γ are also given in Table 9.1 for each iteration. As shown above, Γ is the derivative of the profit function with respect to i^2. This quantity can also be estimated by numerically evaluating $\delta[f(X + \phi*)]/\delta i^2$. This was done by computing $f(X + \phi*)$ for $i^2 = 16.1$. For this value of i^2, $f(X + \phi*) = 2.944026$, and $\delta[f(X + \phi*)]/\delta i^2 = (2.944026 - 2.941090)/(16.1 - 16) = 0.02936$, which is nearly equal to the value for Γ obtained with $i^2 = 16$.

If some cows have records for milk and fat, but not protein, equations [9.92] and [9.93] can then be used to solve for the genetic gain in protein due to selection on milk and fat. G_1 will be equal to the upper-left-hand submatrix of G for milk and fat, and G_2 will be the two first elements of the bottom row of G. The expected genetic gain for protein due to selection on milk and fat is then computed as:

$$\phi_p = 0.01517\phi_m + 0.2873\phi_f \tag{9.101}$$

The partial differentials for milk and fat are then computed as follows:

$$\frac{\delta[F(X + \phi)]}{\delta\phi_m} =$$

$$\frac{(22.9*0.01517 - 0.18)(290 + \phi_f + 285 + 0.01517\phi_m + 0.2873\phi_f) - 0.01517S}{(290 + \phi_f + 285 + 0.01517\phi_m + 0.2873\phi_f)^2} \tag{9.102}$$

Table 9.1. Computation of vector of genetic gains for the optimum selection indices for dairy cattle production traits at two selection intensity levels.

Selection intensity	Iteration number	Genetic gains (kg)[a] milk	fat	protein	Partial derivatives[b] milk	fat	protein	Objective function[c]	Γ[d]
1	1	0	0	0	−0.31	4.04	36.58	1.893	0.1451
	2	92.52	9.05	7.78	−0.30	3.44	35.06	2.177	0.1378
	3	89.51	8.90	7.77	−0.30	3.45	35.07	2.177	0.1379
4	1	0	0	0	−0.31	4.04	36.58	1.893	0.0365
	2	370.13	36.19	31.11	−0.28	1.99	31.11	2.939	0.0293
	3	321.26	33.79	30.88	−0.28	1.99	31.24	2.941	0.0294
	4	321.16	33.79	30.33	−0.28	1.99	31.24	2.941	0.0294

[a] Iteration was on equations [9.85] in the text. Values for genetic gains computed at each iteration were used as initial values for the following iteration. Genetic gains after the last iteration are equal to the initial values for the last iteration.

[b] In Israeli Shekels × 10^{-3} per kg genetic gain.

[c] Profit as estimated from equation [9.93] in Israeli Shekels per kg fat + protein.

[d] Γ = derivative of the objective function with respect to the selection intensity squared.

Table 9.2. Index coefficients for optimum selection indices.

Selection intensity	Genetic gains Absolute milk	fat	protein	Relative milk	fat	protein	Index coefficients milk	fat	protein
1	89.51	8.90	7.77	10.05	1	0.87	−0.141	1	8.37
4	321.16	33.79	30.33	9.50	1	0.90	−0.208	1	12.35

$$\frac{\delta[F(X + \phi)]}{\delta\phi_f} =$$

$$\frac{(4.2+22.9*0.2873)(290+\phi_f+285+0.01517\phi_m+0.2873\phi_f)-1.2873S}{(290 + \phi_f + 285 + 0.01517\phi_m + 0.2873\phi_f)^2} \quad [9.103]$$

Where:

$$S = -0.18(9244 + \phi_m) + 4.2(290 + \phi_f) + 22.9(285 + 0.01517\phi_m +$$

$$0.2873\phi_f) - 5000 \quad [9.104]$$

These values were then used to iterate on equations [9.91]. For i = 4, convergence was obtained after three iterations. Genetic gains at optimum were 1029.57 kg milk, 46.47 kg fat, and 28.97 kg protein. The value of the objective function at the optimum was 2.6964, which is lower than that obtained with the optimum index if all three traits are recorded. At convergence, the partial derivatives were 0.000194 and 0.01124 for milk and fat, and $\Gamma = 0.0226$, which is also lower than for the optimum three-trait index. The index coefficients are 0.0113 and 1 for milk and fat, respectively. This index is only slightly different from direct selection on fat.

9.13 Summary

In this chapter we explained methods used to compute selection indices for nonlinear profit functions. Of the three basic alternatives considered, nonlinear selection indices, restricted indices, and the "graphic method", all have advantages and disadvantages, and none can be considered uniformly "best". Nonlinear indices have been computed only for quadratic and cubic profit functions, and thus are not directly applicable to other profit functions. Furthermore, although these indices select the individuals with the highest breeding values and minimize the mean squared deviation between the index and the aggregate genotype, they do not maximize long-term genetic gain. This problem will be discussed in more detail in the next chapter. Restricted indices, and indices for desired change assume that the long-term objective is known, even if the profit function is nonlinear. This is rarely the case. For more that two traits, index coefficients for the graphic method can only be derived iteratively. Furthermore, by this method, the optimum linear index is dependent on the selection intensity.

Chapter Ten

Comparison of Different Selection Indices

10.1 Introduction

In the previous chapter we considered various alternative selection indices for unknown or nonlinear profit functions. In this chapter we will compare these alternatives, first by theoretical considerations, and then empirically. Wilton, Evans, and Van Vleck (1968) and Ronningen (1971) computed the nonlinear indices that select those individuals with the highest estimated aggregate breeding value. Although *a priori* it would seem that this approach should be optimum, we will demonstrate, following Goddard (1983), that this is generally not the case.

10.2 Linear selection indices for nonlinear profit functions

In the previous chapter we considered cases in which the economic values were functions of the trait values. However different animals in the population under selection will have different trait values. Thus, theoretically the optimum selection index could be different for each animal. To overcome this problem, Wilton, Evans, and Van Vleck (1968) computed economic weights for the population mean for the different traits. This is equivalent to computing profit, P, as the following linear function:

$$P = \frac{\delta P}{\delta x_1} x_1 + \frac{\delta P}{\delta x_2} x_2 + ... + \frac{\delta P}{\delta x_n} \qquad [10.1]$$

Where $\delta P/\delta x_1$, ..., $\delta P/\delta x_n$ are the partial differentials of profit with respect to these n traits computed at the population means of each trait. If the deviation from linearity of the profit function is large relative to the changes in the trait values due to selection, then equation [10.1] will not accurately estimate the change in profit due to selection. This point will be illustrated by the example of Goddard (1983) for the profit function $P = 1/x$. In this example, $\delta P/\delta x = -1/x^2$. Assume that prior to selection $x = 1$, and after selection $x = 1/2$.

From equation [10.1] the change in profit due to selection on x should be: $(\delta P/\delta x)\Omega x$, where Ωx is the change in x. Estimated at the mean of x prior to selection, this value is: $(-1)(-1/2) = 1/2$. However the true change in mean profit will be $1/x_s - 1/x_o$, where x_s and x_o are the mean trait values after and before selection. This value will be $2 - 1 = 1$. This anomaly is illustrated in Figure 10.1.

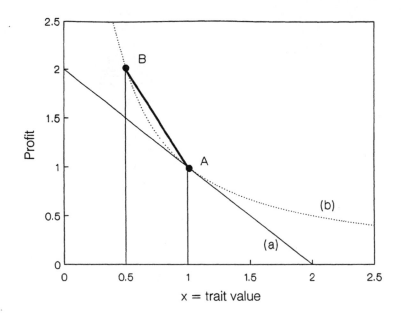

Figure 10.1. Response to selection for an inverse profit function. (b) is the profit function. A is trait value and profit prior to selection, and B is trait value and profit after selection. (a) is the tangent of the profit function at A.

The advantage of the graphic method of Moav and Hill (1966) explained in the previous chapter is that the optimum index is computed as a function of the expected response. This, however, does have the rather undesirable result that the optimum direction of selection will vary with the selection intensity. This is illustrated in Figure 10.2. Two response ellipses are shown for two traits x_1 and x_2. The profit contours for the profit function: $p = x_1 + x_2^2$ are also plotted. As explained in the previous chapter, the optimum selection indices will be at the tangent between the profit contours and the response ellipses. Thus, as shown in this figure, the optimum direction of selection will be different at the two selection intensities.

Furthermore, even if the selection intensity is assumed to be known and fixed, the optimum direction of selection will change over several generations of selection for a nonlinear profit function. In Chapter 8 we considered long-term considerations of selection. Although genetic gains are cumulative, future gains must be discounted more than current gains. In addition most breeding programs will consider only a finite profit horizon. Thus, computation of the economically optimum direction of selection for nonlinear profit functions can be quite complicated.

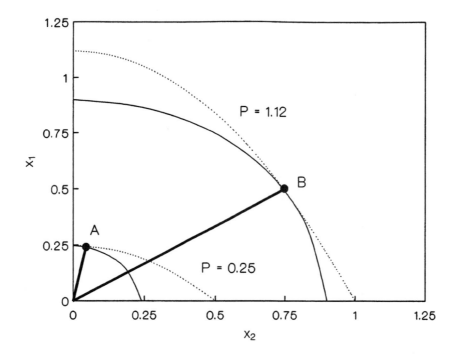

Figure 10.2. Response to selection for a quadratic profit function. Two response ellipses are denoted by solid lines for two traits x_1 and x_2. The profit contours for the profit function: $p = x_1 + x_2^2$ are denoted by dotted lines. A is the point of maximum profit for the small ellipse ($P = 0.25$), and B is the point of maximum profit for the big ellipse ($P = 1.12$).

10.3 Linear vs. nonlinear selection indices

In the previous chapter we computed the optimum quadratic selection index. If profit is a quadratic function of the trait values, then this index will select those individuals with the highest expected aggregate genotype. However, if the profit function is nonlinear, selection of these individuals as parents for the next generation will not lead to the greatest possible gain in the expected mean aggregate genotype of the next generation. Goddard (1983) gives the following example: assume $P = x^2$. If a quadratic index is used, then the individuals with the highest and lowest value for x will be selected. Since the mean of these individuals will be no higher than the population mean, there will be no genetic gain in the next generation with random mating. However, if a linear index is used, then either individuals with high or low values will be selected, but not both, and there will be an increase in the aggregate genotype in the next generation. Thus, in summation, the goal of selection must be to maximize the profit of the mean of the trait values of the selected individuals, and not the mean profit of the selected individuals. Therefore, over the long-term, a linear selection index will always be preferable to a nonlinear selection index.

This conclusion can also be illustrated graphically. As demonstrated in the previous chapter, the possible responses for two traits for a given selection intensity will be described by a response ellipse. By a linear transformation of the trait units, the response ellipse can be converted into a circle. For a given selection intensity, maximum response will be obtained by a linear index. This is, of course, the selection index result for a linear profit function. If there exists a profit contour outside the response circle greater than all profit contours that cross or meet the response circle, then maximum selection intensity (i.e., a linear index) should be applied in the direction of increased profit. If a point of maximum profit exists within the response circle, it is still possible to reach this point by a linear index, but with less than maximum selection intensity. Even in this case, the linear index will be optimum, because decreasing selection intensity saves costs. If less genetic selection is practiced, other criteria can be used for culling of individuals. Furthermore, selection generally requires collecting data on individuals that would not be collected otherwise. If less selection is practiced, then these costs can be reduced.

As an example, consider the case where the population mean is already at the point of maximum profit. Although individuals will differ in their profitability, selection will not increase the mean profitability of the next generation, and can therefore not be justified economically. In practice this situation will occur for traits in which the main profit objective is uniformity. For example the number of days required to hatch eggs, or the time of fruit ripening.

In the previous chapter we considered restricted indices as an alternative to direct selection index methodology. Restricted indices are also linear, but unlike direct selection index methodology or the graphic method do not require that the profit function be known. Under certain circumstances restricted indices can be justified theoretically. Assume the following profit function:

$$P = x_1 - x_2^2 \qquad\qquad [10.2]$$

For a small selection intensity, profit by the graphic method is maximized by increasing x_1 and either by increasing x_2 if its value is negative or decreasing x_2 if its value is positive. However, as selection intensity, or the number of generations of selection increases, the weight given to x_1 in the selection increases at the expense of x_2. This situation is illustrated in Figure 10.3. x_1 and x_2 are assumed to be positively correlated. Extended to infinity, profit gain is maximized by selection on x_1 under the restriction that x_2 is kept constant at zero. This is congruent to computation of a restricted selection index, with the change in x_2 restricted to zero. Thus, over the long-term, a restricted selection index can be justified, even if it does not maximize the gain in profit at each generation.

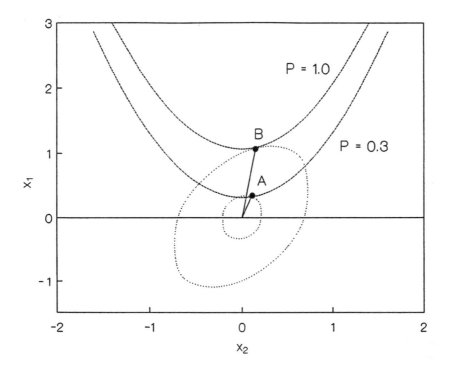

Figure 10.3. Optimum selection index for two response ellipses. Two response ellipses are denoted by dotted lines for two positively correlated traits, x_1 and x_2. The profit contours for the profit function: $p = x_1 - x_2^2$ are denoted by dotted lines. A is the point of maximum profit for the small ellipse (P = 0.3), and B is the point of maximum profit for the big ellipse (P = 1.0).

Until now we have assumed that non-additive variation in both the component traits, and the profit function can be ignored. If the population mean is at the point of maximum profit there may still be considerable non-additive variation in profit. The justification for disregarding non-additive variation is that it is generally not cumulative, and thus will be of little importance over the long-term. Nonlinear indices do have the advantage that they can utilize heterosis on the level of the profit function. This will be considered in more detail in Part 5.

10.4 Suboptimal selection indices, general considerations

We have now discussed at some length the problem of determination of the optimum selection index. We can conclude that in most actual breeding situations the selection index that will be applied will in fact be an educated guess, rather than the theoretically optimum index. The question then naturally arises as to what will be the loss in efficiency due to selection on a suboptimal index. There are four main reasons why a suboptimal selection index may be employed:

1. Use of incorrect values for the genetic and phenotypic parameters.
2. Use of incorrect economic trait values.
3. Exclusion of traits with economic importance from the selection objective.
4. Use of nonlinear selection indices.

Problems in determination of accurate values for the genetic and phenotypic parameters have been dealt with elsewhere at great length, and are beyond the scope of this text. We refer the interested reader to Henderson (1984), although many other publications have dealt with this question. In general we will note that the main difficulty has been to derive accurate estimates for the genetic covariances, especially among several traits. As an extreme example we will note that the literature abounds with reports of genetic correlations outside the parameter space of -1 to 1. Furthermore, if more than two traits are included in the selection objective, it is possible to obtain a matrix of genetic correlations outside the parameter space, even if all the correlations are within the possible range. For a variance matrix to be within the parameter space, all eigenvalues must be non-negative, that is the matrix must be semi-positive definite (Searle, 1982). If at least one of the eigenvalues is negative, then the inverse of the matrix will have a negative value on the diagonal. If selection index is then applied, it would mean that animals that are related would have less similar breeding values than unrelated animals, clearly an insupportable hypothesis. As the number of traits increases, the probability of obtaining a "pseudo-variance matrix" outside the parameter space increases (Hill and Thompson, 1978), unless an estimation method such as multitrait REML (Patterson and Thompson, 1971) is employed, which insures estimates within the parameter space.

We have already dealt at great length in the previous chapters with the problems involved in estimation of the economic trait values. It is therefore quite common to delete traits of economic importance from the selection objective, because their economic values cannot be estimated accurately. This is the common procedure in dairy cattle for most non-production traits.   In Section 3.5 we showed how deleting traits from the selection index affects the efficiency of the index (Cunningham, 1969). From equation [3.29] it is clear that a "reduced" selection index will always be less efficient than the "complete" index, which includes all traits in the aggregate genotype. Furthermore, in some cases negative economic responses were obtained (Gjedrem, 1972).

In the previous chapter we discussed derivation of nonlinear selection indices for nonlinear profit function. In the previous section we demonstrated that nonlinear indices will always be less efficient than linear indices, even if the profit function is nonlinear. Although the quadratic and cubic indices discussed in Chapter 9 are relatively rare in practice, another form of nonlinear index is quite common; threshold selection. That is for certain traits, an economic value of zero is assumed, unless the trait value is below a certain value, in this case the animal is culled. This is commonly done for conformation traits, and also for disease-related traits. It is also practiced for production traits, although in this case it is harder to justify. For example, a cow with fat percent below a certain level will not be used as a bull dam regardless of her performance for other traits.   Threshold selection will be inherently less efficient than linear selection; but, as demonstrated in the previous chapter for the case of a quadratic index, in most cases the difference will be minor.

10.5 Effects of incorrect economic weights on the efficiency of the selection index

Of the four reasons for implementation of suboptimal selection indices, incorrect economic values is probably the most common. At least two studies have shown that relatively small errors in the economic values will have insignificant effects on efficiency of the selection index (Pease *et al.*, 1967; Vandepitte and Hazel, 1977). This is consistent with the result presented in the previous chapter that significant changes in the expected genetic changes due to selection imply very large changes in the economic trait values.

Smith (1983) estimated the relative efficiency of linear indices with incorrect economic values, as opposed to the optimum index over a wide range of possibilities, including economic values up to five-fold the true values and in the opposite direction. From equation [3.28], the relative efficiency, R.S.E., of an alternative index, b*, as opposed to the optimum index was given as follows:

$$RSE = \frac{b'Pb*}{\sqrt{(b'Pb*)(b'Pb*)}}$$ [10.3]

In terms of **a** and **a***, an alternative vector of economic weights, equation [10.3] can be written as follows:

$$RSE = \frac{a'\pi a*}{\sqrt{(a'\pi a*)(a'\pi a*)}}$$ [10.4]

Where $\pi = GP^{-1}G$. This equation is symmetrical in **a** and **a***, so that it is necessary to test only half of the possible parameter space. All traits were standardized to unit phenotypic variance. (This is different from Section 9.8 in which traits were standardized to equal response vectors, following Moav and Hill (1966).) For two traits, pairwise combinations of 2, 1, 0, and -1 for **a** and **a*** were tested, while for more than two traits, all the actual economic values were assumed equal to unity. Heritability was varied from 0.1 to 0.5 for two traits, and from 0.1 to 0.3 for three or more traits. A range of possible genetic correlations, both positive (economically favorable) and negative were tested.

For two traits, the impact of incorrect economic values for a_1 was determined chiefly by the ratio $a_1 h_1^2 / a_2 h_2^2$. Thus if one of the traits dominates the index, incorrect estimation of the economic value of the other trait is unimportant. In general, doubling the economic value for one of the traits decreased selection efficiency by only a few percent. Of course if the sign of the economic value was reversed, the efficiency of the index became zero or negative, unless the effect of the trait was relatively minor. Losses in efficiency are also affected by the genetic and phenotypic correlations, with the genetic correlations having the greater effect.

Similar results were found with more than two traits. That is, if a single trait dominates the index, then changes in the economic values of the other traits will have relatively little effect on the efficiency. Changes are greater when no single trait dominates the index. Loss of efficiency was usually greater with negative genetic correlations among the traits, but in this case the total response possible is low in any event.

Smith (1983) concludes that attempts to achieve precise estimates of the economic weights are not productive, since the gain in selection efficiency will be minimal. This of course is quite important considering the fact that selection of large animals is a long-term procedure. He disagrees with the conclusion of Gjedrem (1972) that all traits of economic importance should be included in the breeding objective. Since the loss in efficiency due to incorrect economic values for traits of minor importance is minimal, they can be assumed to have zero economic value, and therefore be deleted from the selection objective. Decreasing the number of traits for which genetic parameters must be estimated should tend to increase both the probability of obtaining estimates within the parameter space, and the accuracy of the estimates (Hill and Thompson, 1978).

10.6 Summary

If the economic values are nonlinear functions of the traits, then the linear index does not select the individuals with the highest expected mean breeding value. However, for long-term objectives, linear selection indices will always result in greater response to selection than nonlinear indices. If the profit function is nonlinear then the optimum direction of selection will change as a function of genetic improvement. Thus, it is necessary to balance the short-term gain from one generation of selection against the long-term gain of several generations of selection. In general though, changes in mean trait values due to selection will be relatively low within reasonable profit horizons. Furthermore, linear selection index is "robust" to small errors in the economic values. That is, small changes in the economic weights will have practically no effect on the efficiency of selection. However the obverse of this conclusion is that it is necessary to make very large changes in the economic weights to appreciably affect the relative correlated responses of the individual traits.

PART IV

ECONOMIC EVALUATION OF BREEDING PROGRAMS

Compared with the topics of the previous three parts, very little has been written on the economic evaluation of breeding programs, and most of what has been written has dealt with specific examples, rather than general principles. In fact the only paper I have been able to find that deals specifically with this question on a theoretical level is "Economic evaluation of genetic differences," Moav (1973). We will therefore discuss this paper in some detail in Chapter 11, the first chapter of Part IV.

In Chapter 12 we will review the literature sources that have compared alternative breeding programs. Although the literature on this topic is rather extensive, it has mostly dealt merely with comparison of rates of expected or realized genetic progress. Again only a few studies have also considered the relative costs of alternative programs, and we will concentrate on these. In this chapter we will also consider the economic impact of new biotechnology on breeding programs, including embryo transfer and sexing, cloning and selfing of animals, sexed semen, and genetic marker-assisted selection.

In Chapter 13 we will review the literature on evaluation of existing breeding programs. Again this literature has dealt primarily with comparison of realized and expected genetic progress, and much less with other economic factors. In this chapter we will also consider the factors affecting the pricing of

Chapter Eleven

Economic Evaluation of Breeding Programs, Theory

11.1 Introduction

In Chapter 1 we derived equations to estimate expected genetic response to selection on a single trait. In Chapter 3 these equations were extended to cover multitrait selection. In this chapter we will first consider the main cost elements of breeding programs, and then derive equations for the economic evaluation of breeding programs. In Part II we discussed in some length the question of whether the basis for economic evaluation of traits should be profit, economic efficiency, or return on investment. This same question will of course apply to breeding programs. General theory has been developed only in terms of profit, although the criterion of return on investment was also considered by Hill (1971). This question is more acute for the commercial breeder, and will be discussed in some detail in this chapter.

11.2 Major cost elements of breeding programs

Traditionally costs of breeding programs have generally been minimal when compared to the increased income, or efficiency, generated by these programs. It should be noted, though, that many costs that traditionally have been considered part of breeding programs would have accrued in any event, or else generate information which has value beyond the breeding program. For example, the main impetus for milk recording of individual cows was to use this information in progeny tests. However, this information once available is useful to the producer for other farm management decisions. The cost of keeping sires and collecting semen is generally considered part of the cost of the breeding program, even though it would be necessary to keep a minimum number of sires and inseminate females, even if no genetic selection was practiced.

without loss of fertility. This made large scale artificial insemination (AI) economically feasible, and resulted in major increases in the rate of genetic gain for large farm animals (Van Vleck, 1981). Although AI has not had a major impact on the direct costs of breeding programs, other new technologies will. At present multiple ovulation and embryo transplant are becoming economically viable options. In addition, embryo sexing and marker-assisted selection are technologically possible (Lande and Thompson, 1990; Weller and Fernando, 1991; Womack, 1987). Other technologies that have been considered, but are not within current capabilities are semen sexing, cloning, and selfing (Van Vleck, 1981). For the first time, the cost of breeding programs has become a major factor in their economic evaluation. We will discuss now the traditional cost elements, and consider these new factors in the next chapter.

In breeding programs for large animals, recording traits is often the major cost. Although it is now possible to automatically record milk production of each cow, milk samples must be analyzed for component concentration, which is still a relatively costly procedure. Although the main objective of most recording systems is genetic selection, it should be noted that the information recorded also has other uses, such as cow culling, and predicting future production. Certain traits are not included in breeding objectives merely because recording is too expensive. The best example of this is feed consumption for large animals. A question of some importance is whether breeding programs should rely on data recorded by individual producers. This data tends to be less reliable than data collected by professionally trained personnel. Furthermore, the producer sometimes has an economic interest in the values recorded for his own animals. In this case, the data will tend to be biased. This dilemma will be considered in more detail in Chapter 13.

For most large animals, female fertility is very limited, even though many traits of economic importance are only expressed in females. Thus most genetic progress is achieved by progeny testing. A similar situation exists in poultry, in which most selection is based on family, rather than individual selection. Progeny testing can either be performed at regular commercial farms or at specific enterprises dedicated to this goal. In the first case, the cost of progeny testing will be the possible reduction in breeding value by mating to unproven sires, rather than the best sires available, plus an additional factor for risk. It is standard procedure in the US for AI institutes to pay farmers to inseminate cows with semen from unproven bulls. In other countries, farmers are obligated by cooperative agreements to inseminate a fraction of their cows with semen from young sires. In poultry, progeny testing is generally performed at special stations. The commercial producers then buy breeding stock in the form of eggs from the commercial breeder. Often there is an additional stage in which the commercial breeder sells breeding stock to a multiplier who then sell eggs to the general producers.

In species in which the traits of economic importance are expressed chiefly in females, males are maintained only for breeding or for progeny testing. In the absence of genetic selection, it is still necessary to maintain a minimal number of males for breeding, but this number is generally much less than the total number of males that are progeny tested. In the US only about 1/3 of the sires progeny tested are at all returned to service, and the minimum number necessary is actually much lower. In Israel, only one in ten progeny tested sires are returned to service as proven sires. Rather than maintain the males, it is possible in the case of mammals to collect and freeze large quantities of semen over a relatively short period, and slaughter the animals. Thus, the cost of animal maintenance is reduced, but the cost of semen collection and storage is increased. However, since the male may die or become infertile at any time, it can be argued that semen collection during the waiting period is desirable in any event. This alternative is not possible for poultry with present technology; poultry semen loses its fertility after thawing.

Previously, statistical analysis was a non-negligible cost of most breeding programs. However, recent advances in computing equipment have rendered the direct costs of data analysis virtually insignificant relative to other costs. The cost of writing new programs may still be important, but this cost is rarely borne by commercial breeding programs in any event. Over the past several decades statistical methods have become consistently more complex without regard to the increased cost of analysis, even though gains in the accuracy of evaluations have generally been very small (Weller, 1986; Weller, Norman, and Wiggans 1984; Weller, Misztal, and Gianola 1988; Wiggans, Misztal, and Van Vleck, 1988).

11.3 Alternative methods to economically evaluate breeding programs

Similar to the economic evaluation of individual traits, several different methods have been considered to economically evaluate breeding programs. The long-term profit from a breeding program will be a function of the discount rate and profit horizon, in addition to the returns and costs of the breeding program. Thus, one alternative is to assume that the discount rate and profit horizon are fixed, and to compute aggregate profit until the profit horizon is reached (Dekkers and Shook, 1990a; Weller and Ezra, 1989). Alternatively, since gains in the distant future will have a negligible economic value with any reasonable discount rate, some studies have suggested estimating the cumulative costs and returns of one cycle of selection with a fixed discount rate, and the profit horizon set at infinity (Petersen *et al.*, 1974). Since new breeding programs generally require large initial investments, a third alternative is to fix the profit horizon, and estimate the discount rate necessary to achieve a net profit of zero (Hill, 1971). Finally it is possible to fix the discount rate and compute the number of years required to achieve zero net profit (Ferris and Troyer, 1987; Van Vleck, 1981; Van Vleck, 1982; Weller and Ezra, 1989).

We will consider the first alternative in more detail in the next section. In the following chapters specific examples that have used the other alternatives will be discussed.

11.4 Optimization of investment in breeding programs

We will first illustrate how investment can be optimized using the relatively simple formula of Moav (1973), and progress to more complicated and realistic, but less general cases. In equation [4.7] we defined profit, P, as R − C. Returns, R, from a breeding program were computed by Moav (1973) as follows:

$$R = x_D Ma \Delta G \qquad\qquad [11.1]$$

Where x_D is the number of units of produce from each selected animal, M is the number of selected individuals, a is the net present value of a unit of genetic change, and ΔG is the genetic change due to selection for each unit of produce. Thus if a broiler is defined as the unit of produce, then x_D will be the number of chicks produced per selected mother, and ΔG will be measured in terms of broiler weight.

For the simple case of mass selection, we can compute ΔG in equations [1.22] and [1.23] as follows:

$$\Delta G = ih^2\sigma_p/L \qquad\qquad [11.2]$$

where i is the selection intensity, h^2 is the heritability, σ_p is the phenotypic standard deviation and L is the generation interval. In equation [1.21], we showed that $i = z/p$, where z is the ordinate of the normal curve at the point of truncation, and p is the proportion of individuals selected. Thus return from one year of this simple breeding program can be computed as:

$$R = x_D Maih^2\sigma_p/L = x_D Mazh^2\sigma_p/(pL) \qquad\qquad [11.3]$$

In Section 11.2 we listed the major cost elements of breeding programs. We will now assume that the costs of the breeding program can be divided into fixed costs, independent of the number of animals tested, and costs that are proportional to the number of animals tested. With these restraints, costs can be computed as follows:

$$C = K_F + K_A T_n \qquad\qquad [11.4]$$

Where K_F is the fixed costs of the breeding program, independent of the number of animals examined, K_A is costs per animal examined, and T_n is the number of individuals examined. In most traditional breeding programs K_F will consist chiefly of statistical analysis. In a more realistic model, the term $K_A T_n$ should be decomposed into separate terms for males and females. For females the costs will be measuring the economic traits, while for breeding males, costs will consist of raising animals in excess of those necessary merely for reproduction. Since $T_n = M/p$, profit can be computed as:

$$P = x_D T_n azh^2\sigma_p/L - K_F - K_A T_n \qquad [11.5]$$

$x_D ah^2\sigma_p/L$ is annual income per animal selected and per unit of selection intensity. We will define this term as B. Assuming that B is not affected by the breeding program, it can therefore be considered a constant. Substituting into equation [11.5] gives:

$$P = BT_n z - K_F - K_A T_n \qquad [11.6]$$

Thus profit can be expressed in terms of three constants, and two variables z, and T_n. z is of course also dependent on T_n, but this dependency cannot be expressed algebraically. However, the following approximate equality can be used to express i as a function of T_n and M (Smith, 1969):

$$z/p = i = 0.8 + 0.41 \ln [(T_n - M)/M] \qquad [11.7]$$

Substituting equation [11.7] into equation [11.6] gives:

$$P = MB\{0.8 + 0.41 \ln [(T_n - M)/M]\} - K_F - K_A T_n \qquad [11.8]$$

Equation [11.8] is now a function of the same parameters as equation [11.6], with z replaced with a function of T_n and M. In general, M, the number of individuals selected, will be kept to the minimum required by biological, inbreeding, and market considerations. Thus M can also be considered a constant, and maximum profit can be obtained by differentiation with respect to T_n, and equating this differential to zero as follows:

$$\delta P/\delta T_n = 0.41BM/(T_n - M) - K_A = 0 \qquad [11.9]$$

Thus:

$$T_{max} = M(0.41B + K_A)/K_A = 0.41Mx_D ah^2\sigma_p/(K_A L) + M \qquad [11.10]$$

Where T_{max} is the value of T_n for which profit is maximum. Alternatively it is possible to solve for p_{max}, the proportion selected for which profit is maximum, as follows (Hill, 1971):

$$p_{max} = K_A/(0.41B + K_A) \qquad\qquad [11.11]$$

We stress again that this equation is a simplification of any true breeding program, and to the best of our knowledge neither this equation nor a modified form has been applied in practice to determine T_{max}.

11.5 Accounting for differential discounting of costs and returns

We will recall that, on the one hand, genetic changes are cumulative and permanent; but, on the other hand, these changes must be discounted, and gains that accrue after the profit horizon have zero value. In equations [8.5], [8.7], [8,8], [8.9], [8,10] and [8.12] we developed expressions to compute the net present value of a genetic change for successively more complex situations. We will assume that all costs and returns are discounted to the beginning of the breeding program. In the simple program considered above, there is only one product. Thus the cumulative discounted returns can be computed as in equation [8.11]

$$R = V \left[\frac{r^t - r^{T+1}}{(1 - r)^2} - \frac{(T - t + 1)r^{T+1}}{1 - r} \right] \qquad [11.12]$$

with all terms as defined in Chapter 8. In addition to the return from the breeding program, Hill (1971) noted that there is an additional "return", R_i, that can be realized by selling possessions belonging to the breeding enterprise at the termination of the breeding program. (In practice this return is rarely realized, but should be factored into the equation.) Similarly, costs should be divided into initial costs, which need not be discounted, and continuing costs, which should be discounted as given by equation [8.14]. Thus the net present value of the breeding program can be computed as follows:

$$P = V \left[\frac{r^t - r^{T+1}}{(1 - r)^2} - \frac{(T - t + 1)r^{T+1}}{1 - r} \right] - \frac{C_c r(1 - r^T)}{1 - r}$$

$$- C_i + R_i r^T \qquad\qquad [11.13]$$

Where C_c and C_i are the continuous and initial costs, respectively; and the other terms are as defined above. Assuming that $R_i r^T - C_i$ is negligible with respect to the first two terms, this equation can be rewritten as:

$$P = VD_r - C_c D_c \qquad [11.14]$$

Where D_r and D_c are the net present value discounting factors for annual returns and cost respectively. Incorporating these factors, equation [11.8] becomes:

$$P = D_r MB\{0.8 + 0.41 \ln [(T_n - M)/M]\} - D_c(K_F - K_A T) \qquad [11.15]$$

and T_{max} can be computed as:

$$T_{max} = M(0.41BD_r + D_c K_A)/D_c K_A \qquad [11.16]$$

If the profit horizon, T, is extended to infinity, equation [11.13] simplifies as follows:

$$P = \frac{Vr^t}{(1-r)^2} - \frac{C_c r}{(1-r)} - C_i = \frac{R}{d^2(1+d)^{t-2}} - \frac{C_c}{d} - C_i \qquad [11.17]$$

As shown in Chapter 8, and noted by Van Vleck (1982), with discount rates below 0.1, the profit horizon can have a marked effect on the net profit of the breeding program.

In Section 8.4 we considered the dissemination of genetic improvement over a population with overlapping generations, according to the model of Hill (1974). Brascamp (1973) also noted that returns from the four paths of selection occur after different time intervals. For example returns due to selection on the sire-to-dam path occur once the daughter of the selected sire begins milking while returns from the sire-of-sire path only occur after the granddaughters of the selected sires come into milk. Thus returns from the sire-to-sire path should be discounted more heavily, and will have a smaller net present value than returns from the sire-to-dam path. General equations have not been worked out to account for this in optimization of genetic gain.

11.6 Commercial breeders vs. the national interest

Although equation [11.10] will apply both to a commercial breeder and to the whole industry, the specific values of the parameters will be different. On the national level, M which is a function of the national market will be more-or-less fixed. However, a commercial breeder can increase his market share, that is increase M, at the expense of other breeders. Thus the commercial breeder would also be interested in M_{max}, the maximum profit as a function of M, which can be computed by differentiating equation [11.8] with respect to M and equating the differential to zero. In addition, depreciation of genetic gains will be more rapid for the commercial breeder, who must recoup his investment in a relatively short period, as opposed to the national aspect. Furthermore, in a competitive market, the economic value of genetic improvement is likely to be nonlinear. That is if the breeding stock of a particular breeder is below the genetic value of his competitors, it might have close to no economic value, while if the breeding stock is above the level of his competitors, it might have an economic value well in excess of the expected gain to the producer in either profit or economic efficiency. These considerations were considered by Dekkers and Shook (1990, 1990a) for the US dairy industry and will be considered in detail in the following chapter.

11.7 Multitrait breeding programs

As considered in the previous chapter, nearly all breeding programs consider more than one trait. Although in general genetic progress will be maximized by linear selection index as described in Chapter 3, this does not provide a solution as to the economically optimum multitrait breeding program. In addition to the individual economic value of each trait, the different traits may vary as to the time and probability of expression, in which sex the traits are expressed, and the cost of recording for the traits. For example, in dairy cattle milk production is expressed only in females, while beef production is expressed in both sexes. In addition, the main income from beef will be from yearling male calves. Furthermore, milk production is expressed later, but several times during a cow's life, while return for beef production of yearling calves occurs earlier, but only once per individual.

The differential cost of recording various traits is also important. For example, pricing for milk is now generally based on an index of carrier, fat, and protein. It is less expensive to measure milk production than fat, and more expensive to measure protein than either fluid milk or fat. Thus in an optimum breeding program, it is possible that only a fraction of those individuals that are milk-recorded will also be analyzed for fat, and only part of those with fat records will be also assayed for protein. In addition it is possible that in the future, many cows will be assayed for individual milk proteins, as these proteins have differential values in cheese production.

It should be possible to modify equation [11.8] to handle a multitrait breeding program, although in practice this has not been done. The different traits would potentially have different measuring costs, different economic values, and different selection intensities. Instead of a single T_n, there would be a different T_n for each trait. The optimum breeding program, as a function of the number of individuals scored for each trait could then be determined by equating the partial differentials of the profit equation for each trait, and setting each differential equal to zero. It would then be necessary to solve this system of simultaneous equations.

11.8 Summary

In this chapter we considered the major cost elements of traditional breeding programs, which are measuring and recording the traits of interest, progeny testing, maintaining of breeding stock, and statistical analysis. In addition, the development of new biotechnology methods, including multiple ovulation and embryo transplant, embryo sexing, and marker-assisted selection have increased both the potential gains and costs from breeding programs. We explained that different criteria have been used to evaluate breeding programs, and showed how profit could be maximized as a function of the number of animals tested for a simple breeding program based on mass selection. We noted that the considerations leading to the optimal breeding program will be different for national breeding programs, and commercial breeders. Finally we considered briefly the question of economic optimization for a multitrait breeding program.

Chapter Twelve

Comparison of Alternative Breeding Programs

12.1 Introduction

Most of the literature that has compared alternative breeding programs has done so from the aspect of expected rates of genetic gain, generally for a single trait. Only a few studies have attempted to economically compare alternatives, considering both the costs and returns of the different programs, and we will concentrate on these studies. In this chapter we will first summarize the literature that has attempted to economically compare alternative traditional breeding programs. We will then consider the impact of new advances in biotechnology on existing breeding programs. Within this context we will consider both those technologies that are currently available, including multiple ovulation and embryo transplant (MOET), embryo sexing, and genetic marker-assisted selection (MAS); and those technologies which have not as yet been perfected, but may become practical in the future, such as sexed semen, selfing and cloning.

12.2 Half-sib vs. progeny selection for dairy cattle

In most developed countries, dairy cattle sires for use in the general population are selected based on progeny tests (Johansson and Rendel, 1972). That is, young sires are first mated to a limited number of cows to produce between 50 to 200 daughters. Once the production records of these daughters become available, those sires with the highest breeding values are returned for use in the general population, while the remaining sires are culled. The main advantage of this system is that the majority of cows are mated to superior sires with high accuracy evaluations. The major disadvantage is that the generation interval along the sire-to-dam path is much longer than necessary by biological considerations. Semen is first collected from the young sires at the age of one year, but an additional four years will elapse until these sires can be genetically evaluated based on their first crop of daughters.

As an alternative, Owen (1975), suggested a half-sib selection scheme, in which sires of cows are selected based on the performance of their sisters, while sires of sires are selected based on daughter performance. The advantage of this scheme is a major reduction in the generation interval along the sire-to-dam path. The disadvantage is that the accuracy of evaluation is only half of that obtained by progeny tests based on an equal number of production records.

Owen (1975) compared these schemes for a cow population of 100,000, but did not attempt to economically optimize either scheme. Rather he assumed an equal number of 50 sires in use for both schemes. In the progeny test (scheme A) these were divided into 20 proven and 30 young sires, while in scheme B, all sires were "unproven". At equilibrium, genetic gain was 3% higher by scheme A. However, in the first 10 years of the breeding program, cumulative genetic progress by scheme B was four-fold the genetic progress by scheme A. Virtually no genetic progress is generated by scheme A until year ten, when the second crop of daughters from the first group of proven sires are freshened. This is unrealistic for two reasons. First, a breeding program is seldom started completely from "scratch". Rather than use randomly chosen bulls as bull sires, some information is generally available to select better bulls even at the beginning. Alternatively, it is possible to import a small amount of semen of proven bulls from other populations to produce the first crop of young sires. Second, once the first cow records are available, sons of these cows could be used immediately as sires, rather than using a random sample of sires until the first progeny tests are completed, as assumed in this study.

Net costs were 8% less by scheme B, although costs were not discounted, and some of the cost factors were rather arbitrary, for example, a recording incentive of £10 for bull testing per heifer lactation. The main reduction in costs was due to a saving in the cost of keeping bulls during the waiting period in scheme A. In practice this cost could also be reduced by producing semen from these sires during the waiting period, and slaughtering prior to completion of the progeny test.

12.3 Optimization of a dairy cattle breeding program based on progeny test

Ezra and Weller (1989) and Weller and Ezra (1989) optimized a population of 100,000 cows for selection entirely on milk production. They assumed that 20 proven sires were used to inseminate cows not mated to young sires and that six sires were used each year as bull sires. They varied the service period of proven sires, the fraction of cows mated to young sires, and the number of young sires tested. Thus the number of daughters/sire was also varied. Optimum genetic progress was obtained if 40% of the cows were bred to 160 young sires, and proven sires were used for four years. This implies 75 daughters/young sire.

Assuming an initial breeding program in which 15% of cows are bred to

young sires, and 40 sires/year were progeny tested, expected genetic gain would
be 124 kg/year. If the number of sires were increased to 60, genetic gain would
be increased by 2 kg/year to 126 kg/year. This is only 1 kg/year less than the
maximum progress obtainable without increasing the fraction of cows bred to
young sires. With a discount rate of 0.05, an economic value of \$0.2/kg milk,
and a cost of \$4,444 per bull stall, it would take 24 years to reach the break-
even point. By increasing the fraction of the population bred to young sires to
40%, an additional gain of 7 kg/year could be obtained. These results will be
considered again in the following chapter.

12.4   Evaluation of milk and meat production from dual a purpose cattle breed

Hill (1971) considered two breeding schemes for beef production from a dairy
population. He assumed that prior to introduction of a beef breeding program,
a dairy breeding program was already in place. In the first scheme, selection is
for both traits in a single dairy or dual purpose population. In the second
scheme a separate beef nucleus herd is maintained for crossing with the dairy
breed, and progeny of these crosses are sold for slaughter. In scheme one, 600
male calves are first performance tested for growth rate. Of these 150 are then
selected to progeny test for milk production. Since the two traits are not
evaluated simultaneously, the genetic gain will be less than that possible with the
optimum selection index. In scheme two, a nucleus beef herd of 400 cows and
8 bulls is maintained. Selection in this herd is only for beef production. We
will consider separately the costs and returns of each program.

   In scheme one there will be an initial investment for the performance testing
house, which Hill estimated at £50,000. Continuous costs would consist of the
purchase price of the additional 450 male calves to be performance tested each
year. In addition there would be testing costs for all 600 male calves, but part
of these costs would be offset by the slaughter value of the unselected animals.
Hill (1970) estimated the increase in annual costs as £160,000. In scheme two,
the initial costs would consist of purchase of a farm for the nucleus herd, a
testing station for 160 animals/year, and purchase of 16 bulls and 400 cows.
Initial costs for scheme two were estimated at £235,000. However, of this sum,
all but the cost of the testing facility, £12,800, could be realized at any time if
it was decided to terminate the program. The annual costs would consist of
£40/per animal on test, and other minor costs. Thus the total annual cost was
set at £7000/year.

   The total dairy population was assumed to consist of 1,000,000 cows. In
scheme one all cows are mated to sires from the dairy herd. In scheme two,
75% of the cows are mated to dairy sires, and the remaining 25% are bred to
bulls from the nucleus beef herd. All male calves, and all crossbred calves are

finished for beef. In both schemes 500,000 calves are finished each year, of which 400,000 are males. In scheme two, 200,000 of these calves, including nearly all of the 100,000 females are crossbred progeny. Both breeding programs assume a genetic correlation of zero between growth rate and milk production. It is further assumed that selection for growth rate of the bull calves does not affect the rate of genetic gain for milk production in the dairy population.

The expected response to selection in the general population for both traits, and in the nucleus herd for beef production will be a function of selection intensity, the accuracy of the evaluations, and the generation interval. In the dairy population it is assumed that 1/3 of the sires progeny tested are selected as proven sires, and 30% of the cows are mated to young sires. In the nucleus herd 16 bulls are selected each year for mating to the dairy population. Eight of these bulls are also used to maintain the nucleus herd. These bulls are selected from the 160 male calves born each year in the nucleus herd, but the proportion selected is only 0.133, due to 25% wastage. Mean generation interval is 7 years in the dairy population and 2.5 years in the beef nucleus herd.

Once equilibrium genetic gain is reached, gain for growth rate will be 1.26 kg/year in the general population, and 6.08 kg/year in the nucleus herd. Gain in milk production will be equal in both schemes, and is therefore not considered. The unit value of gain in growth rate is assumed to be £0.15/kg. Thus gain from scheme one will be:

$$1.26\text{kg/yr}(£0.15/\text{kg})500,000 = £94,500/\text{yr} \qquad [12.1]$$

Genetic gain for scheme two will be:

$$1.26\text{kg/yr}(£0.15/\text{kg})300,000 + (6.08 + 1.26)0.15(200,000)0.5$$

$$= £166,800/\text{yr} \qquad [12.2]$$

In scheme two, the gain from crossing to the nucleus herd is equal to the mean of the gains obtained in the sire and dam lines.

Three methods were used to economically compare these schemes:

1. Computation of the profit horizon required to reach the break-even point of zero cumulative profit with a fixed discount rate.
2. Computation of the cumulative profit with a fixed discount rate and profit horizon.
3. Computation of the discount rate necessary to reach the break-even point with a fixed profit horizon.

With an 8% discount rate, the break-even point is reached after 15 and 10 years for schemes one and two, respectively. The cumulative profits with a 20 year profit horizon were £1,051,000 and £2,424,000. The discount rates required to achieve a break-even at 20 years were 16% and 27%. These rates can then be compared to alternative investment opportunities. Thus by all three methods, scheme two was superior to scheme one. Unless the initial investment is very substantial as compared to continuous costs and genetic gain, alternative schemes should have the same ranking by all three criteria.

Equation [11.11] was used to compute p_{max}, the optimum selection proportion, and will be repeated here:

$$p_{max} = K_A/(0.41B + K_A) \qquad\qquad [12.3]$$

K_A, costs per individual tested, can be computed as total discounted costs divided by 160, the number of bulls tested per year. Assuming a discount rate of 0.2, total discounted costs for the beef scheme were 4.87(7000) + 235,000 − 0.03(220,000) = £262,490, where 4.87 and 0.03 are the discounting factors for the annual costs and income that can be realized at the end of the program, respectively. Thus, K_A = £262,490/160 = £1641. B, income per animal selected, can be computed as cumulative discounted income divided by the selection intensity and the number of bulls selected. Income from crossing to the beef nucleus herd will be: 6.08kg(£0.15)200,000(0.5) = £91,200. Total income per unit of selection intensity will be £91,200 times the discounting factor and divided by the selection intensity, 1.9. For a discount rate of 0.2, the discounting factor is 7.44. Although 8 sires will be selected, a 25% wastage has been assumed, and the "nominal" number of individuals selected will be 10.67. Thus B = [(7.44)(91,200)]/[(1.9)(10.67)] = £33,470, and:

$$p_{max} = 1641/[1641 + (0.41)33,470] = 0.11 \qquad\qquad [12.4]$$

With a discount rate of 0.08, p_{max} = 0.026. That is, with a lower discount rate, a larger selection intensity, which will result in greater costs for the breeding program, is optimum.

Petersen *et al.* (1974) optimized a breeding program similar to scheme one of Hill (1971). There were two main differences between the two programs. Hill (1971) assumed a finite profit horizon, while Petersen *et al.* (1974) assumed an infinite profit horizon. Hill (1971) assumed that progeny tested sires were kept until the daughter records were evaluated, while Petersen et al. (1974) assumed that a relatively large quantity of semen was collected from each progeny tested sire, and, once this semen was collected, the sire was slaughtered. Thus the cost of semen storage and collection replaced the cost of keeping progeny tested sires until daughter records become available. Petersen *et al.* (1974) assumed a milk-recorded population of 250,000 cows, and selected for kg butterfat production, rather than fluid milk. They varied the number of recruited bull calves for the performance test, selection intensity for growth rate, the number of progeny test daughters per young sire, and the number of stored semen doses per bull. These variables imply that the number of sires progeny tested, and the fraction of the population inseminated with semen of young sires, were also varied.

Except for selection intensity for growth rate, maximum profit for the other variables was obtained at nearly the same value that gave maximum genetic progress for milk production, similar to the results of Ezra and Weller (1989). This was not the case for selection on growth rate, because selection for this trait decreases the selection intensity for milk production. The number of semen doses collected per sire was varied from 10,000 to 50,000, and maximum profit and genetic gain for milk were obtained at the latter value. Thus the optimum may be even higher, although profit was only 0.3 % less with 40,000 doses/sire. At optimum, 40 % of the cows were inseminated with semen from young sires, similar to the results of Ezra and Weller (1989); and 112 sire were progeny tested per year, with 240 daughters per sire. Maximum genetic gain for butterfat was 1.56 % of the mean per year. Seven sires were selected to breed the remaining 60 % of cows, and four sires were selected as bull sires. The fraction of the population inseminated with semen of young sires, and the number of daughters per sire are considerably higher than the breeding programs in most countries.

12.5   Economic impact of multiple ovulation and embryo transplant (MOET) on breeding programs

As noted above several times, female fertility for domestic ruminants is very low, while male fertility, especially via AI, is virtually unlimited. With the advent of MOET in the early 1980's it became technically possible to dramatically increase the rate of female fertility, and thus increase the rate of genetic gain. The cost of MOET, although still high with respect to AI, continues to decrease while success rates increase. Recently techniques have also been developed to determine the sex of embryos prior to transplant (Bondioli *et al.*, 1989), although this technique has not yet come into widespread use. Four main methods have been presented to include MOET in dairy cattle breeding programs:

1. Breeding of all cows in the population by MOET (Van Vleck, 1981, 1982). With this scheme, the cow population in each generation will be the genetic progeny of the best cows in the previous generation.
2. Breeding of bull calves by MOET (Dekkers and Shook, 1990a; Petersen and Hansen, 1977; Ruane, 1988; Weller and Ezra, 1989). By this method it is possible to increase the selection potential on the dam-to-sire path with only a relatively small investment in MOET.

3. Establishment of a nucleus herd based on MOET of all cows (Dekkers and Shook, 1990b; Nicholas and Smith, 1983). Bulls from the nucleus herd are then used to inseminate the general population. In this scheme total MOET costs are still kept relatively low, while the rate of genetic increase in the nucleus herd is greater than in scheme 2.

4. MOET of the best cows in each herd. Cows of inferior genetic merit that would not ordinarily be bred, would be used as foster mothers (Ferris and Troyer, 1987). They assumed that both the costs and gains of MOET would be born by the individual producer, rather than the national or regional breeding program.

All of these studies have considered only breeding for a single trait, usually milk or some simple index based on milk components. We will now consider these schemes in detail. Scheme 1 gives the greatest genetic increase, but has by far the greatest cost. Van Vleck (1982) assumed that the 10% best cows were used as embryo donors, all cows and sires were produced by ET, at a cost of $300 per live birth via ET, an economic value of $0.15/kg milk, and a zero interest rate. Under these conditions, genetic gain was increased by 35 kg/yr, but cumulative profit only became positive after 100 years. With embryo sexing, only half as many transfers would be necessary, and a positive net profit would be obtained after 50 years, assuming that embryo sexing did not increase the cost of the procedure. With any realistic discount rate and profit, schemes of this type are therefore not economical, unless there is a dramatic reduction in the cost of MOET.

Contrary to the situation with scheme one, all of the studies that have considered scheme two have found it to be economically feasible, under certain conditions. Weller and Ezra (1989) assumed that with MOET the number of bull dams could be decreased by a factor of three. They further assumed a population of 100,000 cows, a discount rate of 5%, an economic value of $0.2 per kg milk, an initial cost of building a MOET facility of $300,000, and an annual cost of $150,000 to perform 300 transplants/year, i.e., $500/transplant. Genetic gain was increased by 7 kg/year (compared to a base of 124 kg/year without MOET). A cumulative zero profit was obtained after 16 years, and after 20 years cumulative profit was $2,029,000. As shown above in Chapter 8, a relatively small genetic gain can have a large effect on returns and profit. Dekkers and Shook (1990a) considered the additional revenue that would accrue from AI for bull dams in a situation of several competing AI firms, but did not consider the costs of MOET. They found that the increase in genetic gain would be 0.015 trait standard deviation units. With a standard deviation of about 1000 kg for milk, this is twice the gain predicted by Weller and Ezra (1989), but they assumed a more than six-fold reduction in the number of bull dams selected.

A MOET nucleus herd was first proposed and discussed in detail by Nicholas and Smith (1983), but they only considered the additional genetic progress that could be obtained. To date no study has attempted to economically

.... . , . 'gram of this type, also considering the additional costs.   The advantage of the nucleus herd scheme is that MOET costs are kept relatively low, because they are only performed within the small nucleus population. Genetic gain in the nucleus herd is transferred to the general population by insemination with bulls produced in the nucleus herd. The disadvantages are as follows:

1. A more-or-less self-contained breeding program must be planned for the nucleus herd. Although it will be possible to achieve greater rates of genetic progress along the dam-to-dam path via MOET, rates of progress will be less along the other three paths of inheritance, due to the small population size.
2. Intense selection in a small population will lead to significant increases in inbreeding, which has a negative effect on most traits of economic importance, such as milk production and growth rate (Hudson and Van Vleck, 1983; Weller, Quaas, and Brinks, 1990).
3. When cows in the general population are inseminated with bulls from the nucleus herd, only half of the genetic gain obtained in the nucleus herd is transferred to the general population (Hill, 1971).

Nicholas and Smith (1983) considered two selection schemes for the nucleus herd; a *juvenile* and an *adult* scheme. In the juvenile scheme, male calves are selected for breeding, and female calves are selected for MOET at the age of one year based on their pedigree. In the adult scheme, cows are selected for MOET and bulls are selected for breeding at the age of three, with cow selection based chiefly on their own first production record, and bull selection based on the records of their dam and sisters. The juvenile scheme has the advantage that generation interval is shortened, but the accuracy of the evaluations is lower. Rates of genetic gain and inbreeding increase with the number of progeny per female donor, and the number of donors per male. Rate of gain is slightly greater with the juvenile scheme, but inbreeding levels are significantly higher. A genetic gain of 0.129 s.d. units/year was obtained with an adult scheme of 512 cows, and 1024 embryo transfers/year. This is about 30% greater than can be obtained with a traditional progeny testing scheme. However, only half of this gain will be transferred to the general population by insemination with bulls from the nucleus herd.

Within a herd, Ferris and Troyer (1987) found that break-even costs for ET ranged from $300 to $500 per ET cow entering the herd, with a profit horizon of 20 years, provided that the donors were among the 5% best cows in the herd, and were mated to superior sires. An economic value of $0.13/kg milk and a discount rate of 0.1 were assumed. Increasing the number of ET's per donor dam from 2 to 4 reduced the break-even cost by 20%. The break-even cost declined with increase in the selection intensity of donor dams. That is, the less ET is performed in a herd, the more economical it becomes. In conclusion, at current prices, it may be economically feasible from the point of view of the individual farmer, to perform MOET on the one or two best cows in his herd.

12.6 Economic impact of embryo sexing and splitting, cloning, semen sexing, and selfing on breeding programs

The main impact of embryo sexing would be to decrease the number of embryo transfers necessary in a breeding program that includes MOET. For schemes that utilize MOET on the dam-to-sire path, the number of transfers could be reduced by half, so embryo sexing could be economically viable if the cost of sex determination on a single embryo is less than the cost of a single ET. For a MOET nucleus herd, in which both males and females are produced by ET, the gain would be less, and the number of transfers could be reduced 30%.

Via embryo splitting it is possible to produce genetically identical individuals similar to identical twins. Four identical calves from embryos split twice have already been produced (Womack, 1987), although this technique is still in the experimental stage. The routine use of embryo splitting in conjunction with MOET would result in an increase in the accuracy of genetic evaluation, because it would be possible to pool the records of several genetically identical individuals. In conjunction with a MOET nucleus breeding scheme, embryo splitting could increase the rate of genetic gain by up to 10% (Nicholas and Smith, 1983), without increasing the inbreeding level. Further increases in genetic gain could be obtained, but would result in an increase in the inbreeding level.

"Cloning" is the production of large numbers of genetically identical individuals. Cloning of both embryos and mature individuals has been considered, although neither is possible with current technology. To some readers cloning of mature individuals may seem far-fetched, but in fact from a breeding point of view, this is equivalent to vegetative reproduction, which is quite common in many plant species. If cloning of mature individuals becomes possible, then animal breeding could become quite similar to plant breeding, in which commercial production consists of raising large numbers of genetically identical individuals, with genetic breeding programs performed only at special institutes on relatively small numbers of individuals. Of course it is not possible to economically evaluate a technique beyond current technology. Even cloning of embryos would result in a significant increase in the rate of genetic gain above that possible with embryo splitting and MOET (Nicholas and Smith, 1983; Van Vleck, 1981).

Although semen sexing has been considered a distinct possibility for at least 20 years (Foote and Miller, 1971; Soller and Bar-Anan, 1973), and reports of successes have been published from time to time, no reliable method has at present been developed to separate "X" and "Y" sperm without significant loss of fertility. With sexed semen it would be possible to dramatically increase the selection intensity along the dam-to-dam path, and marginally increase the selection intensity along the dam-to-sire path. Rate of genetic gain would be increased by 15 kg/year (Van Vleck, 1981). Assuming that the proportion of cows selected as dams of dams is 0.9 under normal circumstances, and 0.45 with sexed semen, the expected genetic gain of a daughter of a selected dam will be 36 kg without sexed semen, and 163 kg with sexed semen. With a discount rate of 0.1, a profit horizon of 15 years, and an economic value of $0.11/kg milk, the net present value of mating to a selected dam to produce a heifer would be $14 in the standard progeny test program, and $63 for a program with sexed semen. Van Vleck (1981) assumed that six semen doses are required to produce a heifer in a standard breeding program. Thus at $10 per semen dose, the net loss incurred in the production of a milking heifer is $60 - $14 = $46. With sexed semen, only half the cows selected as cow dams would be mated with semen from proven sires. The remaining cows, who would be mated with "male" semen could be inseminated with inexpensive semen. Assuming that the inexpensive semen is priced at $7/dose, while semen from elite sires is priced at $10/dose, the break-even point for the cost of semen sexing would be $19/dose.

Soller and Bar-Anan (1973) noted that sexed semen would reduce the economic value of beef traits in the dairy herd, since the designated dams of male calves could be bred instead to a beef breed. Instead, selection for calving ease would become more important, as breeding dairy cows to large beef breeds increases the frequency of dystocia.

Another breeding technique that is common in plants, but beyond current biotechnology for vertebrates is "selfing". That is, producing progeny with the same individual as both the male and female parents. This technique could be useful to reduce the number of daughters required for a progeny test. For example, the records of 12 daughters produced by selfing of young sires would give equivalent accuracy of evaluation to 50 normally bred daughters. Selfing of sires would of course entail embryo transplant, and this must also be factored into the potential cost. A major drawback of selfing is that the progeny are 50% inbred. As stated above, even relatively minor levels of inbreeding have a significant negative effect on production traits.

12.7 Genetic marker-assisted selection

Traditionally animal breeding has been chiefly trait-based selection, without regard to the specific Mendelian loci that contribute to the breeding value of the selected individuals. A number of studies have shown that individual loci that affect quantitative traits can be located with the aid of genetic markers. Until recently large numbers of segregating genetic markers were not available in domestic animal populations. With the advent of restriction fragment length polymorphisms (RFLP), (Soller and Beckmann, 1982) and more recently DNA microsatellite polymorphisms (Litt and Luty, 1989), this situation is changing. Of all the biotechnology techniques considered, only semen sexing and marker-assisted selection (MAS) do not require ET.

A number of studies have considered the expected genetic gain possible with MAS. In conjunction with traditional breeding programs, MAS can be used either to decrease the generation interval, or to increase the accuracy of the evaluation. For the latter alternative, MAS will be useful chiefly for low heritability traits (Smith and Simpson, 1986; Lande and Thompson, 1990). For high heritability traits, little is gained if the genotypes of individuals with respect to individual quantitative trait loci are known. For sex-linked traits the main advantage of MAS could be to shorten generation intervals. Kashi *et al.* (1986) suggested a two stage selection procedure for young sires; screening of male calves for QTL genotype, followed by a progeny test of those calves selected in the first stage. They estimated that the rate of genetic gain could be increased by 25 to 50% of traditional progeny test breeding programs.

Another potential of MAS is to select for non-additive genetic effects, which are generally ignored in traditional breeding programs. Finally, MAS may also be more efficient for multitrait selection, especially with negative genetic correlations among the traits in the selection objective (Lande and Thompson, 1990; Weller and Fernando, 1991). Development of a MAS program in dairy cattle will probably require genotyping about 10,000 individuals for at least 15 markers each (Weller, Kashi, and Soller, 1990). Thus, assuming a conservative cost of $1 per genotype determination per locus per individual, the total cost of genotype determination will be $150,000, which is of a similar magnitude to the additional costs required for incorporating MOET of bull dams into a progeny selection breeding program.

12.8 Summary

Similar to most other economic enterprises, genetic improvement is subject to the law of diminishing returns, that is the first investment in breeding produces substantial returns, but as investment increases, marginal gains decrease. Only a few examples of optimizations of breeding programs are presented in the literature, and these tend to be highly simplified. Three different methods have been used to economically compare alternative breeding programs; cumulative profit, number of years to reach the break-even point, and the interest rate that results in zero profit for a fixed profit horizon; but results were generally similar. Until the advent of modern biotechnology, costs of nearly all breeding programs were relatively low. Thus, for single trait selection, the economically optimum breeding program was nearly equal to the breeding program that gave optimal genetic progress.

With the advent of new technologies, based chiefly on MOET, which are still relatively expensive, this situation is changing, and the economically optimal breeding program may be significantly different from the breeding program that gives maximum genetic gain. Economic considerations, such as interest rate and profit horizon also become more important. Of course it is not possible to economically evaluate options beyond current technology, such as sexed semen, or selfing, but similar to the case of MOET, it should be remembered that the mere fact that a procedure is technologically possible, does not mean that it is economically feasible.

Chapter Thirteen

Evaluation of Existing Breeding Programs

13.1 Introduction

Very little has been directly written on economic evaluation of ongoing breeding programs. The literature that does exist deals chiefly with estimation of realized rates of genetic gains, and comparison to the theoretical values for optimum genetic progress. Realized genetic progress has generally been considerably less than the theoretical expectation. We will attempt to explain this result within an economic context. In free-market economies, the price of breeding stock is usually set by supply and demand. We will review those studies that have considered the market price of breeding stock, and compare actual market prices to the expected economic value of genetic improvement as developed in Chapters 6, 7 and 8. Although we have maintained that maximum progress is achieved by a linear selection index, we will see that the actual market price of breeding stock is in many cases a nonlinear function of the trait breeding values.

13.2 Comparison of realized and predicted rates of genetic progress

Estimation of genetic trends for most domestic animal species is complicated by the fact that management conditions also change over time. Thus genetic trend cannot be estimated merely by comparing production levels over time. To obtain reliable estimates of genetic trends requires long-term recording with overlapping generations, or maintenance of control populations not under selection. For many commercial populations neither of these options is available. Most estimates of long-term genetic trends in commercial populations have been for the major traits under selection, especially milk production in dairy cattle. General policy has been to estimate the regression of sire breeding values, weighted by the number of daughters per bull, on the birth or freshening date of the daughters. Different methods, such as the regression of cow breeding value on birth year, tend to give significantly different results. It is much more difficult to obtain meaningful genetic trends for other species with less advanced recording systems, or for secondary traits in dairy cattle, although a few estimates do exist, and will be considered.

Estimates of genetic trends for milk production in the US, Canada, and New Zealand were summarized by Van Vleck (1986). Lee, Freeman, and Johnson (1985) found that genetic trends in the US population from 1960-1969 were only 1.55 and 2.55 kg/year, based on cow and bull evaluations, respectively. During the period 1969-1978, genetic trends were 52.55 and 83.73 kg/year, for the cow and bull evaluations. Powell and Norman (1985) found similar results for the same population. Although the genetic trends during the later period are much higher, they are less than the expected value of about 100 kg/year, which is still a conservative estimate. A number of studies considered in the previous chapter have shown that rates of genetic gain in the range of 0.1 standard deviation units are well within the capability of advance progeny testing programs. The phenotypic standard deviation for milk production is within the range of 1,200 to 1,400 kg.

Weller, Ron, and Bar-Anan (1986) and Weller and Ron (1989) estimated annual genetic trends for the Israeli dairy cattle population for both primary and secondary trends from 1976 through 1988. Trends for production traits were 89 kg milk, 2.2 kg fat and −0.008% fat for 1976 through 1983, and 147 kg milk, 2.45 kg fat and −0.025% fat for 1978-1988. Genetic trends were economically positive for fertility and calf mortality, but negative for dystocia. There was very little intended selection for these secondary traits during the periods considered.

13.3   The reasons for genetic trends being less than the theoretical maximums

From the examples given above, it can be concluded that in general, genetic trends have been less than the theoretical expectations. From the previous chapter, it should be clear that the breeding program that gives maximum genetic gain is not necessarily the program that gives maximum net profit. Furthermore, even if the goal is to maximize profit, the question is whose profit? In the previous chapter, breeding programs were economically evaluated from the national or regional aspect. The situation may be quite different when it is necessary to evaluate a breeding program separately from the point of view of the farmer, the commercial breeder, and the national interest.

The individual farmer, who is the direct consumer of genetic improvement, is not able to independently evaluate the genetic potential of the breeding stock that he purchases. This will be especially true for breeding stock of similar genetic merit. Moav (1973) defined this problem in terms of a "range of non-discrimination." That is, farmers are not able to discriminate between breeding stock that differ only slightly in genetic merit. Thus, within the range of non-discrimination, it is in the breeders' interest to decrease breeding costs as much

as possible. Assuming that this will also lead to a decrease in genetic merit, it
is possible to obtain a negative correlation between breeders' profit and genetic
merit over the range of non-discrimination. Furthermore, for a commercial
breeder, it may be more worthwhile to invest in advertising, which is visible, than
in increasing genetic gain, which is not. In most countries the range of non-
discrimination is kept to a minimum by independent evaluation of competing
breeding stock, either by government agencies or by universities. In both cases
the assumption is that the institution that performs the evaluation will not be
subject to influence by commercial breeders who have a stake in the results of the
evaluation.

   Another reason for less than optimal genetic progress is that costs that may
be zero or negligible from the national aspect may be quite substantial from the
point of view of the commercial breeder. For example, in the US farmers are
paid to have their cows inseminated with semen from young, unproven sires.
From the national point of view, insemination of cows with semen of young sires
does not increase the cost of the breeding program. Inseminating cows with
semen from young sires requires no extra effort than inseminating cows with
semen from proven sires. However, this "cost" must be born by the commercial
breeder, and passed on to his consumers. It also means that a breeder can
increase his profit by decreasing the number of inseminations from young sires.
Thus, the breeding program that maximizes the profit of the commercial breeder
can be radically different from the program that would maximize national genetic
gain, or even maximize national net profit.

   This will be illustrated with the following example based on the US dairy
cattle population. Equation [11.8] can be modified to compute the optimum
selection proportion of young sires and the optimum number of daughters per sire
as follows:

$$P = MB\{0.8 + 0.41 \ln [(T_n - M)/M]\} - K_F - K_A T_n \qquad [13.1]$$

with all terms as defined in Chapter 11. We will first modify the term $K_A T_n$ to
include an additional term for fixed costs per daughter sampled. Dividing by M,
profit per bull selected, P_M, is computed as follows:

$$P/M = P_M = B\{0.8 + 0.41 \ln [(R_M - 1)\} - K_F - K_A R_M - K_s n_s R_M \qquad [13.2]$$

where K_s is fixed costs per daughter sampled, n_s = number of daughters sampled
per sire, and $R_M = 1/p$. B, the value of the genetic gain per year per animal
selected per unit of selection intensity, is computed as follows:

$$B = 0.5 x_d \sigma_g A_c a \qquad [13.3]$$

Where x_d is the number of semen doses sold per sire, σ_g = 600 kg milk is the
genetic standard deviation, A_c is the accuracy of the evaluation, and a = $0.05

is the net present value of semen from a sire with a breeding value of an additional kg of milk. Assuming $h^2 = 0.25$, A_c will be approximately equal to $[n_s/(n_s + 15]^{1/2}$ (Van Vleck, 1981). Note that for progeny evaluation $h^2\sigma_P$ of equation [11.5] has been replaced with $\sigma_g A_c$. The factor 0.5 is included because the daughters receive only half of their genes from their sires. Assuming that $n = 20,000$; substituting equation [13.3] into equation [13.2] gives:

$$P_M =$$

$$\frac{300,000\sqrt{n_s}\ [0.8 + 0.41\ \ln\ [R_M - 1)]}{\sqrt{n_s} + 15} - K1 - K_A R_M - K_s n_s R_M \qquad [13.4]$$

To find the optimum values for R_M and n_s it is necessary to differentiate this equation with respect to these variables, and equate to zero, as follows:

$$\frac{\delta P_M}{\delta R_M} = \frac{123,000\sqrt{n_s}}{(R_M-1)\sqrt{n_s} + 15} - K_A - K_s n_s = 0 \qquad [13.5]$$

$$R_M = \frac{123,000\sqrt{n_s}}{(K_A + K_s n_s)\sqrt{n_s} + 15} + 1 \qquad [13.6]$$

$$\frac{\delta P_M}{\delta n_s} = \frac{2,250,000}{\sqrt{n_s}(n_s + 15)^{3/2}}\ [0.8 + 0.41\ \ln\ [R_M - 1)] - K_s R_M = 0 \qquad [13.7]$$

$$0.8 + 0.41\ \ln\ \left[\frac{123,000\sqrt{n_s}}{(K_A + K_s n_s)\sqrt{n_s} + 15} \right] =$$

$$= \frac{0.0547 K_s n_s (n_s + 15)}{K_A + K_s n_s} + \frac{K_s\sqrt{n_s}(n_s + 15)^{3/2}}{2,250,000} \qquad [13.8]$$

Equation [13.8] can be used to solve for n_s, and then equation [13.6] can be used to solve for R_M for any given values for K_A and K_S. Note that n_s and R_M are independent of K_F. We will consider two cases: a cooperative and a commercial AI institute. For both cases we will assume that the net present cost of keeping a sire until his daughter proofs become available is $10,000. In the first case there will be no additional cost per sire progeny tested, and $K_A = $10,000$. In the second case we will assume that the stud must pay an additional mean purchase price of $10,000 per bull calf, so that for the commercial AI institute, K_A = $20,000. For the cooperative AI institute we will assume that the cost per milk-recorded daughter, K_S, is a nominal $20, while the commercial AI institute pays the farmer a total of $100 per milk-recorded daughter. With these values, R_{max} = 10.3 and 5.2 for the cooperative and commercial AI institutes, respectively. The corresponding values for n_s are 126 and 66 daughters, respectively, which correspond to accuracies of 0.94 and 0.90. The values for the commercial AI institute for both variables are considerably less than the values that give either maximum profit from the national point of view, or maximum genetic gain. They are however not very different from the actual mean values for the US dairy industry.

In the US Holstein population, about 1000 young sires are progeny tested per year (Miller, 1988), and about 300 are returned to service as proven bulls. Although the selection proportion is in fact greater than 0.3, because the better proven sires are used more intensively, it is still much less than the optimum levels estimated in the previous chapter. Van Tassell and Van Vleck (1987) found that the realized selection differential over the last available five year period was only 28% of the theoretical possible value. The number of daughters per young sire is about 50, thus only 50,000 are bred to young sires. The total US cow population is about 10,800,000 cows (Gruebele, 1988). Assuming a culling rate of about 25%, this means that 2,500,000 heifers come into milk each year. Thus only about 2% of new heifers are milk-recorded progeny of young sires. These numbers can be compared to the optimum levels of about 40% found by both Ezra and Weller (1989) and Petersen *et al.* (1974), and the realized value of 15% for the Israeli dairy population (Weller, 1988).

Lower than necessary selection intensities result also from the pricing system of semen common in most countries. Generally there is a very large price differential between the semen of the most elite sires, and second tier sires, and this price differential is greater than the actual economic value of breeding to the elite sires (Wilder and Van Vleck, 1988). Farmers are aware of this fact, and therefore elect to inseminate most cows with semen from less expensive sires. Generally only elite cows with the potential to be bull dams are mated with the most expensive semen. The price of the semen of the elite sires is reduced when sires with even higher evaluations become available.

In addition to the factors considered above, Van Vleck (1986) found that longer than necessary generation intervals; selection emphasis on non-production traits, especially conformation; and biases in genetic evaluations, also reduce the actual rate of genetic gain. Van Tassel and Van Vleck (1987) found that the sum of generation intervals along the four paths was 30 years, while the required minimum, considering biological and breeding limitations, was only 25 years. The sire-to-sire and dam-to-sire paths were 3 and 2 years greater than necessary, respectively.

In both the US and Canada, a relatively large emphasis has been put on conformation traits in sire selection (Van Vleck, 1987).   Standard practice has been to set minimum levels for fat percent and conformation traits in the selection of bull dams.   As shown in Chapter 3, any alternative form of selection will be less efficient than selection index.   Furthermore, truncation selection of this type can result in an unexpectedly large emphasis on the traits selected by truncation.   This problem was considered in general terms by Moav (1973).   He noted that the implied economic weights of the farmer may often be different from the economic weights in the national interest.   His example was for production and reproduction in swine.   The individual farmer may be more interested in litter size, because this increases the number of pigs he can sell. However, from the national point of view it may be more important to increase growth rate.   In a commercial breeding situation, it is the farmer who will determine the relative economic value of different traits.   We will consider emphasis on conformation again in the following section.

A main problem in genetic evaluation has been preferential treatment of potential bull dams.   In most developed countries, bull studs will pay large prices for male calves of elite cows.   This has led farmers to try to make good cows seem better than they really are.   A number of studies have indicated that the actual regressions of the evaluations of AI bulls on their dams are less than the predicted values (Van Vleck, 1986).

It may be of interest that, in the Israeli population, in which emphasis on conformation has been less than in the US, the price paid for bull calves is only slighter higher than the beef price, and semen of all sires are priced equally; realized genetic gain has been close to the theoretical values.

13.4   Actual pricing of breeding stock vs. theoretical considerations

Wilder and Van Vleck (1988) compared the price of semen of 324 US Holsteins to the sire evaluations for production and conformation traits.   The effect of number of daughters was also included.   They analyzed three AI institutes separately and 11 institutes jointly.   They also repeated the analyses excluding sires with semen prices greater than $100/dose.   The coefficient of determination for the complete data set was only 0.26, and ranged from 0.37 to 0.57 for the individual studs.   Eliminating the four "outliers" increased R^2 to 0.48 for the combined analysis, and from 0.44 to 0.70 for the individual studs.   Inclusion of quadratic effects for evaluations for conformation and production increased R^2 significantly, especially for the joint analysis.   This indicates that farmers tend to prefer "balanced" sires with good evaluations for both conformation and production, as opposed to sires with superior evaluations for some traits, and low evaluations for others.   This preference is of course contrary to our conclusion in Chapter 10, that a linear selection index will always result in maximum genetic progress. Even with quadratic effects, the highest R^2 for a single stud was only 0.75. The availability of semen is also probably a significant factor in price determination, but was not included as a factor in the analysis.

In the US sire summary, sires are ranked chiefly by the total performance index, (TPI). Until the late 1980's, TPI was a linear index of the sires' evaluation for milk, percent fat, and conformation. Wilder and Van Vleck (1988) compared the partial regressions of the sire evaluations for these traits on semen price to the economic values implied by TPI, again excluding the sires with extremely high semen prices. In general the calculated partial regressions were similar to the implied economic values, except that slightly more emphasis was put on conformation than implied by the TPI. When the outliers were included in the regression analysis, emphasis on conformation was even greater. The partial regression of milk on semen price was \$0.03/kg. This value is surprisingly low. Most of the studies considered in the previous chapter have assumed a discounted net value in the range of \$0.05 to \$0.1/kg. Van Vleck (1981) assumed a value of \$0.055 for a profit horizon of 10 years, while McGilliard (1978) found that the net present value of semen was about one quarter of the nominal milk price, which in the US has been recently about \$0.25/kg. This implies an economic value of \$0.06/kg, or twice the regression found by Wilder and Van Vleck (1988).

The TPI was constructed so that the economic values for milk, percent fat, and conformation, per standard deviation unit would be in the ratio of 3:1:1. If the pricing of semen reflects farmer demand, then their emphasis on conformation is even greater than the TPI value. This would conform to the conclusion of Van Vleck (1986) given above, that emphasis on conformation has been a major cause in the lower than expected genetic trends for milk production.

The non-linear nature of farmer demand from the commercial breeder was also noted by Shultz (1986) for the case of poultry. He observed that most traits will have a minimal acceptable level from the point of view of the farmer or consumer. If a breeding stock is below this level, then it will be nearly unsalable, while above this break-point, all values will have equal economic merit. The commercial breeder will therefore concentrate on those traits in which he is weakest, and ignore those traits in which he is above his competitors. Selection of this type will also lead to less than optimum genetic progress.

13.5 Summary

Although classical economic theory assumes that the public interest is generally best served by free competition, this is not necessarily the case for genetic improvement. In many cases the conflicting interests of breeders, farmers, and consumers will lead to suboptimal levels of genetic progress. This will be especially true if the suppliers of breeding stock are independent commercial enterprises. In this case, costs that might be negligible from the national view, can be very significant for the commercial breeder. Since it is nearly impossible for the buyer to independently evaluate breeding stock at the time of purchase, misrepresentation is a major problem. Although many industries have adopted independent evaluation procedures, the possibilities for "beating" the system are still great. For these reasons, realized genetic progress has been lower than the economically optimal level from the national point of view.

PART V

CROSSBREEDING AND HETEROSIS

Until now we have considered economic aspects of genetic improvement within a single breed. Although selection within a single breed is the rule for certain objectives, especially dairy cattle, many production systems make use of crossbreeding between different breeds at some stage of the production process. The main justification for crossbreeding strains has been to obtain heterosis, although there is considerable confusion as to the exact meaning of this term. The term "heterosis" was coined by G. H. Shull in 1914 (Shull, 1948) to include hybrid vigor as the manifest effect of a "developmental stimulation resulting from the union of [genetically] different gametes." Although heterosis is generally defined as superiority of the hybrid over both parents (Strickberger, 1969), numerous studies, particularly those involving beef cattle, have estimated heterosis as any deviation from the mean of the parental strains.

The "classical" explanations for heterosis are elimination of inbreeding depression, and overdominance at the level of the individual locus. We will see that even in the absence of these "true" genetic effects, crossbreeding is often more profitable than selection within a single line. Moav (1966) defined five types of "economic" heterosis, and these will be explained in Chapter 14. In Chapter 15 we will explain criteria to determine the most profitable parental combination, and in Chapter 16 we will show how planned matings together with line breeding can be economically optimized. Unlike genetic selection, the effect of heterosis is not permanent and cumulative. Thus as shown in Section 8.3 the net present value of a gain from crossbreeding will be much less than an equal

Chapter Fourteen

Economic Evaluation of Heterosis

14.1 Introduction

This chapter will be based chiefly on Moav (1966) and Allaire (1977). The remaining chapters in Part V will consider three other papers in this series (Moav, 1966a; Moav, 1966b; and Moav and Hill, 1966) in detail. We will utilize methodologies developed in Chapter 6 to estimate the economic value of traits from profit equations. We will also use the graphic method of Moav developed in Chapters 6 and 9, in which genetic values are superimposed on a map of profit contours. We will demonstrate how "economic heterosis" can be achieved even if the individual traits are additive on the scale of measurement. Five different types of economic heterosis will be defined. We will further show how planned matings within a breed can increase the mean economic value of the offspring, provided that for at least one of the traits the economic value is nonlinear on the economic scale.

14.2 Profit equations with separate sire and dam lines

In Chapter 6 we first considered the relationship between the effects of reproductivity and productivity on profit of poultry and swine production. For the example of an integrated enterprise that raises both mother hens and layers, profit, P, was expressed by equation [6.20] which we will repeat here:

$$P = K - V_2 - V_1 \qquad [14.1]$$

Where K is return per unit production less fixed costs per unit production, V_2 represents the variable costs of production, and V_1 represents the variable costs of reproduction. We also noted in Chapter 6 that profit contours in this case will be curved lines. This point will be considered again below.

Three genotypes directly determine the profit of the enterprise, the genotype of the "commercial offspring", its dam, and its sire. Each of these genotypes can potentially affect the three terms in equation [14.1] that determine profit. Thus from a genetic point of view, profit can be expressed as follows:

$$P = S(g_s) + D(g_d) + O(g_o) \qquad [14.2]$$

Where S, D, and O are the effects of the sire, dam and offspring on profit as a function of their genotypes, and g_s, g_d, and g_o, are the three genotypes. Each genotype can potentially have a different effect on the components of profit in equation [14.1]. In this context we are only considering the direct effects of each individual to profitability, not effects due to genetic correlations. This will be illustrated with a few examples.

In dairy cattle the male has a direct effect only on male fertility, while the female has a direct effect on female fertility, calving ease and milk production traits. However, since the number of males kept for breeding is a minuscule fraction of the number of females that are raised for production, only the genotypes for female production and female fertility are of major economic interest. Thus males are selected based on their breeding values for these traits, rather than their phenotypes for male fertility, which can be measured directly. For egg production in broilers, the commercial broiler is killed before reaching reproductive age, and a single sire services at least ten hens. Since the level of egg production does not seem to affect growth rate or meat quality, only the contribution of the dam to reproduction costs is important. For egg production in laying hens, the reproduction rate of the dam, and her daughter, the commercial layer, are both economically significant. Again reproduction costs of the sire will be negligible, and other considerations will be used to select the optimum sire line.

The genotype of the offspring can be expressed as $\frac{1}{2}(g_s + g_d + g_{sd})$, where g_{sd} is the deviation of the genotype of the offspring from the mean of the parents. In the absence of heterosis, g_{sd} will consist only of Mendelian sampling, and will have an expectation of zero. We will now consider a case in which the direct contribution of the sire and dam to production, and the direct contribution of the offspring to reproduction are negligible. That is, production will be a function of the genotype of the offspring, and reproduction will be a function of the dam. Assuming income as constant, profit will then be equal to:

$$P = K - (1/2)O(x_{2s} + x_{2d} + x_{2sd}) - D(x_{1d}) \qquad [14.3]$$

Where O is the production costs as a function of the offspring's genotype, D is the reproduction costs as a function of the dam's genotype, and x_{2s}, x_{2d}, and x_{2sd} are the sire, dam, and sire X dam interaction genotypes for the production trait, and x_{1d} is the dam's genotype for the reproduction trait.

In Chapter 6, we first considered a case where profit was a linear function of the production trait, and an inverse function of the reproductive trait. This relationship for the example of pig production was expressed in equations [6.21], [7.9] and [9.75], which we will repeat here:

$$P_1 = K_1 - K_2x_2 - K_3/x_1 \qquad [14.4]$$

Where P_1 is profit per pig marketed, x_1 is number of pigs weaned per sow per year, x_2 is age to a fixed market weight, K_1 is income less costs independent of x_1 and x_2, K_2 is costs dependent on x_2, and K_3 are fixed costs (feed and non-feed) per sow.

Of course the same general relationship will also hold for broiler production in poultry. Moav and Moav (1966) estimated the values of the constants in equation [14.4] for the situation in the British poultry industry in 1966. Inserting these values into equation [14.4] gives:

$$P_1 = 10.6 - 0.1x_2 - 320/x_1 \qquad [14.5]$$

Where P_1 is profit in pence per pound live-weight of broiler, x_2 is days from hatching to slaughter at a fixed weight of 3.8 lbs, and x_1 is egg production per hen. For the remainder of this section we will assume complete heritability, so that the phenotypic value is equal to the breeding value of each trait.

14.3 Graphic representation of heterosis

In Chapter 6 we showed how equation [14.5] could be used to construct a map of "profit contours" in which all points on a contour have an equal level of profit. The profit contours for this function are shown in Figure 14.1. As in Chapter 6, the profit contours are curved. A specific cross between two lines is shown superimposed on the profit contours. Genetic additivity is assumed for both traits on the scales measured, thus the genotype of the offspring is at the midpoint between the two parental lines. However, the profitability of the cross will be a function of both the offspring's and dams' phenotypes, as given in equation [14.4]. The profitability of the cross is marked as the point SD, and is at a higher profit contour than either, S, D, or O. Intuitively this result is actually quite obvious to most animal breeders. If two strains are available, one that excels at reproduction, and a second that excels at production, it is advantageous to use the former as the female parent, and the latter as the male parent. Alternatively profit can be increased by using a smaller strain as the female parent and a larger strain as the male parent. Thus K_3 is decreased, and x_2 is increased. This is common practice for beef cattle in many production systems.

Note also that although complete genetic additivity was assumed, the value of the profit contour at O is not equal to the mean of the profit contours at S and D. This is due to the fact that the profit contours on the scales of measurement used are not straight lines. The question of the scale of measurement is generally not a problem for breeding within a line, since in this case ranking of individuals will not be affected by changes in the scale. However, even for a single line, most models of selection and genetic evaluation assume additivity.

For crossbreeding, this question takes on special significance. In Figure 14.2 the same parental combination is plotted on the same profit contours, but in this case genetic additivity is assumed on the scale of $1/x_2$ and x_1. Therefore the y-axis is no longer linear in x_2, and the difference between 60 and 80 days is shown of equal scale to the difference between 60 and 48 days. In this case the profit value of the offspring is higher than either parent, even though both traits are genetically additive. This anomaly will be considered in more detail in the following section.

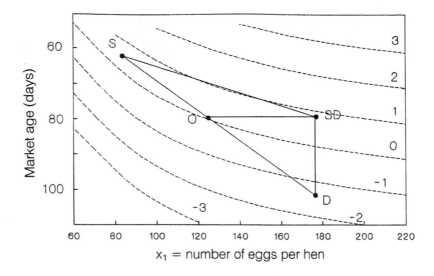

Figure 14.1 Profit of a cross between two strains of broilers. Broken lines are profit contours. S is the profit value of the sire strain. D is the profit value of the dam strain. O is the profit value of the offspring. SD is the profit value of the cross.

The question then arises as to the optimum scale for trait measurement. Wright (1952) suggested that the "best" scale is the one in which the effects of the various factors of interest are additive. In breeding we would be interested in a scale that yields genetic additivity. This means that in a cross between two strains, the mean of the F-2 would be near the mean of the parental lines, and that the F-2 would have a normal distribution. This would not necessarily be the case for the F-1, due to overdominance and interactions at specific loci. For the specific question at hand, Moav (1966) suggested the following additional criteria: 1) linearity and independence in the profit equation, 2) minimization of the magnitude of risks from operational decisions when assuming additivity on a scale which is genetically non-additive, and 3) ease of measurement.

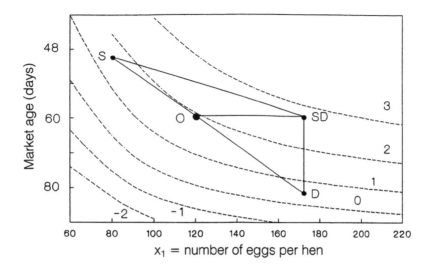

Figure 14.2 Profit of a cross between two strains of broilers with additivity on the inverse of market age. Broken lines are profit contours. S is the profit value of the sire strain. D is the profit value of the dam strain. O is the profit value of the offspring. SD is the profit value of the cross.

In the example given above, complete linearity of profit is obtained only if the traits are additive on the scales of x_2 and $1/x_1$. However, this is the least realistic scale with respect to x_1. On this scale the egg production of the crossbreed would be at the harmonic mean of the parental strains, that is $2x_{1s}x_{1d}/(x_{1s} + x_{1d})$, where x_{1s} and x_{1d} are the egg production of the sire and dam lines, respectively. Since egg production in both strains is greater than unity, the expected production of the crossbreed, under this assumption would always be less than or equal to the mean of the parental lines. In practice the opposite is generally true. In general the location of the point O will be more susceptible to scale effects than the point SD. It should be emphasized that this problem is only important when the range of strain values is large with respect to the mean of all strains. If differences between strains are small relative to the mean value, deviations from linearity will be negligible.

We will define "profit heterosis" as the deviation of the profitability of a specific sire-dam cross from the midpoint of the two parental strains. It should now be clear that even with genetic additivity on the level of the individual traits, there can be significant profit heterosis for a specific sire-dam cross, and the economic value of the cross will be different if the sire and dam lines are reversed. Furthermore, even with additivity at the level of the component traits, two pairs of strains with equal midpoint values can have different profit values for a cross. This situation is illustrated in Figure 14.3 for two crosses between four lines of poultry. Although the midpoint of the parental lines is the same for each cross, the values of the crosses, denoted by the points SD and S_1D_1 are different. SD has a higher profit value even though the profit values of the parental strains, S and D, are lower than those of the parental strains for the alternative cross. Thus, it is not sufficient merely to evaluate the parental lines in order to determine the most profitable cross.

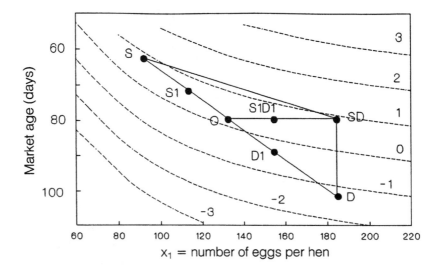

Figure 14.3. Profit of two crosses among four strains of broilers. Broken lines are profit contours. S and S_1 are the profit values of the sire strains. D and D_1 are the profit values of the dam strains. O is the profit value of the offspring from both crosses. SD is the profit value of the S-D cross. S_1D_1 is the profit value of the S_1-D_1 cross.

14.4 The five types of profit heterosis

Moav (1966) divided profit heterosis into five distinct effects.

   1. Heterosis of component traits. This can consist either of overdominance or interactions at the level of individual loci. To estimate this effect it is necessary to measure the trait on a scale in which it is genetically additive, or to

estimate the effects of individual loci, which is much more difficult. In plant species in which many strains of a species can be readily crossed, this effect is termed "specific combining ability" to differentiate it from "general combining ability" which is equivalent to an estimate of the additive breeding value of the strain. Only a few examples of overdominance at the level of individual loci for trait of economic importance have actually been presented in the literature, and only for plant species (Weller, Soller, and Brody, 1988).

2. Heterosis due to sex-linkage. In mammalian species the male has one X and one Y chromosome, while the female has two X chromosomes. Many important traits have been localized to the X chromosome, which is much larger than the Y chromosome. A male passes his X chromosome to his daughters and the Y chromosome to his sons. Thus there is no genetic similarity between a sire and his sons for chromosome X-linked traits. This can result in a sex-linked difference among the progeny of specific crosses.

3. Maternal effects. For most domestic species, the contribution of the sire to the offspring is only half of the nuclear genetic material. In addition to chromosomes, the female has three additional effects on the offspring: cytoplasmic inheritance, a prenatal maternal effect, and a postnatal maternal effect. Mitochondria also contain DNA. This cytoplasmic genetic material is passed by the egg cell directly from mother to progeny without reduction at meiosis. Conflicting reports have been presented as to the importance of cytoplasmic inheritance in milk production traits (O'Neill and Van Vleck, 1988). In all domestic species the dam will have a prenatal effect on the offspring, and in certain production systems a significant postnatal effect as well. For mammalian species this will consist chiefly of the dam's milk production. Weller, Brinks, and Quaas (1990) found that the dam genetic effect on weaning weight in dairy cattle was greater than the direct genetic effect.

4. "Nonlinearity" heterosis. If the component traits are additive on a scale that is nonlinear to profit, then the profit value of the offspring will be different from the midpoint of the parental profit values. The difference between this effect and the first type of heterosis is that the first type would be considered only if the component traits were additive on the scale of profit. This will be illustrated using the example of equation [14.4]. Assuming complete heritability, the parental mean of two lines for profitability, P_m, can be computed as follows:

$$P_m = (P_s + P_d)/2 = K_1 - K_2(x_{2s} + x_{2d})/2 - K_3(x_{1s} + x_{1d})/(2x_{1s}x_{1d}) \qquad [14.6]$$

Where P_s and P_d are the sire and dam profit values, respectively. Assuming additivity on the scale of measurement, the profit value of the offspring, P_o will be:

$$P_o = K_1 - K_2(x_{2s} + x_{2d})/2 - 2K_3/(x_{1s} + x_{1d}) \qquad [14.7]$$

The difference between P_o and P_m is a measure of the nonlinearity heterosis, H_{nl}, and can be computed as follows:

$$H_{nl} = P_o - P_m = \frac{K_3(x_{1d} - x_{1s})^2}{2x_{1s}x_{1d}(x_{1s} + x_{1d})} \qquad [14.8]$$

In this case the nonlinearity heterosis is due only to a difference in the last term of the right-hand side of equations [14.6] and [14.7].

  5. Sire-dam heterosis. This is the most important component of heterosis, and the main justification for crossbreeding. This will be illustrated again using equation [14.4]. In practice one line will be used as the dam line and the other line as the sire line. Assuming additivity on the scale of measurement and complete heritability, the profitability of the sire-dam combination, P_{sd}, will be:

$$P_{sd} = K_1 - K_2(x_{2s} + x_{2d})/2 - K_3/x_{1d} \qquad [14.9]$$

The magnitude of the sire-dam heterosis, H_{sd}, can be computed as follows:

$$H_{sd} = P_{sd} - P_m = K_3(x_{1d} - x_{1s})/(2x_{1s}x_{1d}) \qquad [14.10]$$

The ratio of sire-dam to nonlinearity heterosis is then computed as follows:

$$\frac{H_{sd}}{H_{nl}} = \frac{x_{1s} + x_{1d}}{x_{1s} - x_{1d}} \qquad [14.11]$$

Thus the sire-dam heterosis will be greater than the nonlinearity heterosis if x_{1d} is greater than x_{1s}. That is, if the dam line has higher fertility than the sire line. This of course will be the general situation.

14.5 Planned matings within a single line

For dairy cattle nearly all breeding schemes are based on selection within a single line. Even so, a fair amount has been written, especially in the non-scientific press about the advantages of crosses between specific sire and dam pairs. In the dairy industry the economic advantage obtained by a specific sire-dam combination above the midparent breeding values is termed "nicking". Most results presented on the advantage of nicking have been anecdotal. That is, a particular farmer reports that a specific sire-dam mating gives good progeny. In dairy cattle, with few offspring per female it is very difficult to accurately estimate non-additive components of variance. Part of the reason for the emphasis on nicking by farmers is that it has virtually no cost. A farmer must mate his cows to a given list of available sires. Why not plan the matings so as to maximize the profitability of the sire-dam combinations?

It is generally assumed that mating of opposites for specific traits is desirable, that is, a cow that is exceptional for production, but with only mediocre conformation should be mated to a sire with opposite characteristics. Assuming additivity of all component traits on the profit scale, there will be no over all gain in profitability if this procedure is followed.

Allaire (1977) noted that this kind of disassortative mating would reduce both phenotypic and genetic variance. Since uniformity is often itself a goal of animal production systems, it is possible that the mating scheme described above can be justified on this basis. However, Allaire (1977) further noted that in practice it is unlikely that disassortative planned mating could of itself lead to a significant reduction in genetic variance due to the relatively low heritability of most economic traits, and incomplete phenotypic correlations among traits. Furthermore, long-term laboratory experiments with disassortative mating have not produced positive results.

In the previous sections we noted that the effect of female fertility on profit will be nonlinear. This will be true for most other non-production traits, such as disease resistance and conformation traits. In Chapter 9 we discussed nonlinear selection indices, and in Section 9.3 we developed the optimum quadratic selection index for situations in which the aggregate genotype is a quadratic function of the trait values. For two traits, the aggregate quadratic genotype, H_q will have the form:

$$H_q = a_1 g_1 + a_2 g_2 + a_3 g_1^2 + a_4 g_2^2 + a_5 g_1 g_2 \qquad [14.12]$$

Where a_1 through a_5 are the economic values, and g_1 and g_2 are the breeding values for the two traits. The optimum quadratic index, I_q, was given in equation [9.19], and for two traits, x_1 and x_2, will have the following form:

$$I_q = b_1 x_1 + b_2 x_2 + b_3 x_1^2 + b_4 x_2^2 + b_5 x_1 x_2 \qquad [14.13]$$

Where b_1 through b_5 are index coefficients. If profit is a quadratic function of a trait that is genetically additive on the scale of measurement, then the economic value for a specific sire-dam mating will have a similar form, with x_1 and x_2 replaced with the sire and dam values, respectively. Wilton, Evans, and Van Vleck (1968) demonstrated for a quadratic profit function that if selection is based on breeding values estimated for each trait, then the selection index coefficients ($b_1, ..., b_5$) will be equal to the economic values ($a_1, ..., a_5$). The same principle holds true for the optimum sire-dam mating index. That is:

$$I_m = a_1 y_s + a_2 y_d + a_3 y_s^2 + a_4 y_d^2 + a_5 y_s y_d \qquad [14.14]$$

Where I_m is the optimum mating index and y_s and y_d are the estimated sire and dam breeding values, respectively.

Allaire (1977) gives the following example. Assume that profit is a quadratic function of a trait x, as follows:

$$P = 60x - 10x^2 \qquad [14.15]$$

The profit of individuals with trait values 1, 2, 3, 4, and 5 will be 50, 80, 90, 80, and 50, respectively. Assuming genetic additivity on the scale of x, and complete heritability, the profit of a specific sire-dam combination will be:

$$P = 60(x_s + x_d)/2 - 10(x_s + x_d)^2/4 \qquad [14.16]$$

$$P = 30g_s + 30g_d - 2.5g_s^2 - 2.5g_d^2 - 5g_sg_d \qquad [14.17]$$

Where x_s and x_d are the sire and dam values for x. Changing the sire–dam combinations will affect the overall progeny mean for profit. For example assume two sires and two dams. One sire and dam have trait values of 1, while the other sire and dam have trait values of 5. If sires and dams are mated assortatively (low to low and high to high) then the mean offspring profit value will be 50. However if the two pairs are mated disassortatively (low to high) then the mean offspring profit value will be 90. Based on the results of Wilton, Evans, and Van Vleck (1968), with incomplete heritability, the trait values can be replaced with the estimated breeding values and the optimum mating index for this profit function will be:

$$I_m = 30y_s + 30y_d - 2.5y_s^2 - 2.5y_d^2 - 5y_sy_d \qquad [14.18]$$

Where y_s and y_d are the sire and dam breeding values for x. This will be the case only if all traits of economic importance are expressed only in the progeny. For the example in equation [14.4] the profit of the cross was determined by the dam's phenotypic value for the fertility trait.

Extension to the multitrait case is straightforward and not conceptually different from a cross between two lines considered in the previous sections. If at least one trait is nonlinear to profit on the scale of genetic additivity then nonlinearity heterosis can be exploited by planning specific sire-dam matings. Allaire (1977) considered in detail a two-trait situation, consisting of a production trait linear in profit, and a maintenance trait where returns diminish greatly from a maximum level for trait values below a threshold value. Above the threshold all values for the maintenance trait have equal profitability. On a practical level one could consider an example such as egg shell thickness. A certain minimum trait value is considered "acceptable". Values above this threshold do not increase profit, while below the threshold profit declines dramatically because eggs with shell thickness below the threshold value are virtually unsaleable. Again in this case disassortative mating will result in a greater mean offspring profit value than random mating.

14.6 Summary

"Heterosis" is generally defined as a situation in which the trait value for the progeny of a cross is greater than the trait value of either parent. "Economic heterosis" or "profit heterosis" is in fact a combination of five separate effects, each of which can generate heterosis on the scale of economic value. The main component of economic heterosis was defined by Moav (1966) as "sire-dam" heterosis. To achieve a net sire-dam heterotic effect over the population, it is generally necessary to maintain separate sire and dam lines. Nonlinearity heterosis can be obtained by planned matings within a line, and therefore adds virtually no additional cost to the breeding program. Unlike selection for additive genetic variance, increased economic value obtained by heterosis is not cumulative or permanent. As shown in Section 8.3, with a profit horizon of ten years and a discount rate of 0.15, the net present value of an increase in profit due to heterosis will only be one sixth of the cumulative economic value of an equal nominal profit gain from additive genetic selection. This calculation is based on the assumption that the same heterotic effects are generated each year to the profit horizon.

Chapter Fifteen

Choice of the Most Profitable Parental Combination

15.1 Introduction

If a number of different lines are available for crossbreeding then the choice of the most profitable parental combination can be quite complex. In this chapter we will continue the example considered in the previous chapter of one production trait and one reproduction trait. We will first consider the situation in which the component traits are genetically additive on the scale of measurement, and then consider the situation in which the component traits are not genetically additive. Again this chapter will be based largely on the graphic method of profit representation developed by Moav.

15.2 The choice of the most profitable parental combination when only two parental lines are available and traits are genetically additive

In equation [14.4] profit per broiler was computed as a function of a production trait, growth rate, and a reproduction trait, the number of eggs per hen. We will repeat this equation with minor modifications:

$$P = K_1 - K_2 y - K_3/x \qquad [15.1]$$

For simplification x_2 was replaced with y, and the subscripts were removed from P and x_1. The profit contours were plotted for this function in Figure 14.1 using the values for K_1, K_2, and K_3 of equation [14.5]. The profit contours are curved because profit is an inverse function of egg production. Assuming complete heritability and genetic additivity, the trait values of the offspring will be equal to the parental midpoints for both traits. As in the previous chapter we will consider only the case in which the profitability of the sire-dam cross is greater than the profitability obtained by either line separately. This requires that $x_d > x_s$, and $y_d < y_s$, where x_d and x_s are the egg production of the dam and of the dam and sire lines, respectively; and y_d and y_s are the growth rates of the dam and sire lines, respectively.   The trait values for a specific cross between two lines are shown in Figure 15.1 on a map of profit contours. The points S, D, and O denote the trait values of the sire, dam, and offspring, and the point SD is the profit value of the cross. As explained in the previous chapter, the profit value for SD will be higher than either parent or the offspring.

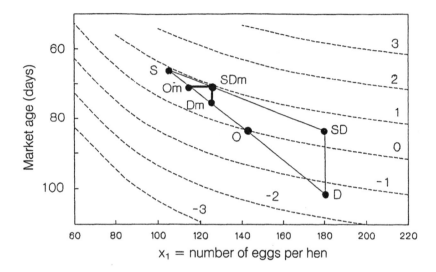

Figure 15.1. The most profitable cross between two strains of broilers. Broken lines are profit contours. Points S, D, and O denote the trait values of the sire, dam, and offspring, and the point SD is the profit value of the cross. SO is the profit value of the cross between S and O. D_m is the optimum female parent. O_m is the profit value of the offspring of S and D_m. SD_m is the profit value of the optimum cross.

If the offspring is mated back to the sire, then a backcross progeny is produced with trait values at the midpoint of S and O. However, as in the previous case, egg production will be determined by the dam, which in this case is O. Thus the profitability of this cross will be at the point SO. Note that this point is on the straight line connecting S and SD. The slope of this line will be $(y_d\text{-}y_s)/[2(x_d\text{-}x_s)]$. Replacing y_d and x_d with y_o and x_o, which are the parental midpoints, will not change the slope of the line.   Since the sire contributes to profit only through the production of the offspring, maximum profit will be obtained when the strain with the highest production is used as the male parent of the commercial cross. However, the economically optimum male parent of the commercial cross. However, the economically optimum dam line may be a cross between the two parental strains. By making a series of crosses and backcrosses between these strains, with S as the sire line for the final cross, it is possible to obtain the profit value of any point on the line connecting S and SD. Since the profit contours are curved, the profit values of these points will not be equal. As shown in Section 9.8 the point of greatest obtainable profit will be the point of tangency between the profit contours and the line S-SD. Any lower profit contour will cross this line at two points, while any higher profit contour will not meet the S-SD line.

As in Section 9.8 the point of tangency can be computed by finding the profit contour with a slope equal to the slope of the S-SD line for a point on this line. Starting from equation [15.1], y as a function of P and x is computed as follows:

$$y = (K_1 - P)/K_2 - K_3/(K_2 x) \qquad [15.2]$$

The tangent of the profit contours is then computed as the differential of equation [15.2] with respect to x as follows:

$$\tan P(x) = \delta y/\delta x = K_3/(K_2 x^2) \qquad [15.3]$$

Equating this tangent to the slope of the S-SD line gives:

$$K_3/(K_2 x_m^2) = (y_d - y_s)/[2(x_d - x_s)] \qquad [15.4]$$

Where x_m is the x-value for the dam of the most profitable cross. Solving for x_m gives:

$$x_m = \left[\frac{2K_3(x_d - x_s)}{K_2(y_d - y_s)} \right]^{1/2} \qquad [15.5]$$

The economically optimum dam line will be denoted D_m. The trait values for the cross between S, the optimum sire line, and D_m will be the means of S and D_m for both traits. This point will be denoted O_m. The profit value for this cross will be at SD_m, the point of tangency between the S-SD line and the profit contours. D_m, O_m and SD_m are also marked on Figure 15.1. Defining y_m as the value of the offspring for y, the point O_m, with coordinates (x_m, y_m) will also lie on the S-D line. Thus the value for y_m can be also be found by equating slopes as follows:

$$(y_d - y_m)/(x_d - x_m) = (y_d - y_s)/(x_d - x_s) \qquad [15.6]$$

$$y_m = (x_m y_s - x_d y_s - x_m y_d + x_s y_d)/(x_s - x_d) \qquad [15.7]$$

It is possible that x_m may be outside the range of x_d-x_s. If $x_m > x_d$ then a simple cross between S and D with D as the female parent will result in higher profit than any other cross between the two lines. Conversely if $x_m < x_s$ then the sire line will have a higher profit value than any cross between these lines.

15.3 Maximum profit when three parental lines are available and traits are genetically additive

With more than two possible parental lines the possible crosses increase exponentially. For example, with three lines, it is possible to preform nine basic crosses; each line as the sire combined with each of the three lines as the dam. In the next generation, it is possible to cross the F-1 either as the sire or the dam with either one of the original three lines, or with a cross between them. How then can the best possible combination be determined from the myriad of possibilities? We will first determine principles for the case of three lines using the graphic method for illustration.

From the discussion above it should be clear that certain combinations will be more profitable than others. Thus if line 1 has both higher production *and* fertility than line 2, then line 1 will be superior to any cross between these lines. (At this point we are still assuming complete heritability and genetic additivity.) Thus the only case of interest is when one line has higher fertility while the other line has higher production. Since the sire only affects production, *a priori*, the line with the highest production will be chosen as the sire line. The x and y values for three potential parental lines are plotted in Figure 15.2. The line with the highest production is denoted S, while the other two lines are denoted D_1 and D_2. D_1 has a higher reproductive rate than D_2, while D_2 has a higher production rate than D_1. Thus D_2 is intermediate for both traits.

If D_2 is located below the line S-D_1 then no cross involving D_2 can have a higher profit value than the best cross involving progeny of S and D_1. Conversely, if D_2 is located above the S-SD_1 line, then no cross involving all three lines can have a higher profit value than the best cross between either S and D_2 or D_1 and D_2. However, if the slope of the S-D_2 line is between the slopes of the S-D_1 and S-SD_1 lines, it is possible to obtain a higher profit by crossing the three lines than that obtainable with any combination of only two lines. The line connecting S, SD_2 and SD_1 is the line of maximum profit obtainable by crosses between these three lines. The point of tangency between this line and the profit contours will be the point of maximum profit, and is denoted P_{max} on Figure 15.2. Note that in this case P_{max} is above both the S-D_1 and S-D_2 lines. Thus P_{max} can only be achieved by a three-way cross between these lines. In this example a maximum profit is obtained by first crossing D_1

and D_2, and then mating this F-1 as the female parent with S as the male parent.

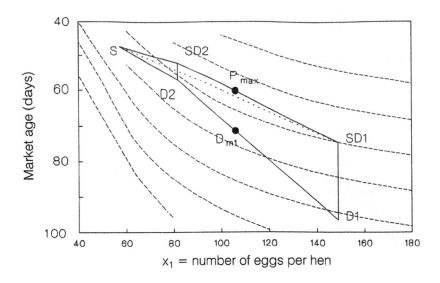

Figure 15.2. The most profitable cross between three strains of broilers. Broken lines are profit contours. S is the profit value of the sire strain. D_1 and D_2 are the profit values of the three dam lines. SD_1 and SD_2 are the profit values of the crosses between S and D_1 and D_2. D_{m1} is the female parent of the optimum cross. P_{max} is the profit value of the optimum cross.

Even if the conditions described above are met, it is still possible that a two-way cross is most profitable. If the line connecting S, SD_2, and SD_1 is tangential to the profit contour in the segment S, SD_2 then a cross involving only S and D_2 will be more profitable than any combination of all three lines. The method of equating tangents can again be used to algebraically determine the cross of maximum profit. An x-value for maximum profit can be found for both the S-SD_2 and SD_2-SD_1 segments. We will define x_m as the x-value for maximum profit for the S-SD_2 segment. This point is derived by equation [15.4] with the values for x_d and y_d replaced with the appropriate values for the D_2 strain. The slope of the SD_2-SD_1 segment will be $[y_{d1} - (y_s + y_{d2})/2]/[2(x_{d1} - x_{d2})]$. The x-value for maximum profit along the SD_2-SD_1 segment, x_{m1}, can be found by substituting the appropriate values in equation [15.5] as follows:

$$x_{m1} = \left[\frac{2K_3(x_{d1} - x_{d2})}{K_2[y_{d1} - (y_s + y_{d2})/2]} \right]^{1/2} \qquad [15.8]$$

If $x_{ml} > x_m$ then a cross of all three lines will result in more profit than any cross of S and D_2. Since the profit contours are convex while the S-SD2-SD1 broken line is concave, $x_{ml} > x_m$ implies that $x_m = x_{d2}$. The rules for determining the most profitable combination of three parental lines are summarized in Table 15.1.

As in the case of two lines, the strain with the highest production will always be the optimum male parent for the final cross. The optimum female parent for the commercial cross can be any of the three lines, or a cross between S and D_2, or D_1 and D_2. If the optimal female parent is one of the original lines, then P_{max} will have the fertility value of optimal female strain. If the optimal female parent is a cross between S and D_2 then P_{max} will be located at SD_m, and only these two lines will be used to produce the commercial broiler. Only in the fourth situation are all three lines used to produce the maximum profit cross. However, even in this case, it is not necessary to maintain all three lines, since it should be possible to maintain the cross between D_1 and D_2 as a "synthetic" line.

Table 15.1. Rules to determine the economically optimum cross among three parental strains.

Situation	Location of P_{max}	Maximum profit
$x_m < x_s$	S	$P_S = K_1 - K_2 y_s - K_3/x_s$
$x_s < x_m < x_{d2}$	SD_m	$P_{SDm} = K_1 - K_2(y_s + y_m)/2 - K_3/x_m$
$x_m < x_{d2}$ and $x_{m1} < x_{d2}$	SD_2	$P_{SD2} = K_1 - K_2(y_s + y_{d2})/2 - K_3/x_{d2}$
$x_{d2} < x_{m1} < x_{d1}$	SD_{m1}	$P_{SDm1} = K_1 - K_2(y_s + y_{m1})/2 - K_3/x_{m1}$
$x_{m1} > x_{d1}$	SD_1	$P_{SD1} = K_1 - K_2(y_s + y_{d1})/2 - K_3/x_{d1}$

15.4 Maximum profit when many parental lines are available, and traits are genetically additive

In the previous section we demonstrated that when three potential parental strains are available, a cross involving all three strains will be most profitable only under rather restrictive conditions. These conditions will now be made general for the case of many strains, illustrated in Figure 15.3. The x and y values for a number of potential parental strains are located on a map of profit contours. Obviously any strain with both lower fertility and a lower production level than an alternative strain can be excluded from consideration as a potential parental strain. Thus, starting with the strain with the most favorable value for the production trait we will first consider the strain with next best production, provided that it has higher fertility than the previous strain. Following this principle strains can be added until the strain with the highest fertility is reached. These strains can then be connected with a broken line that will have a negative slope in all segments. From the previous section it should also be clear that if any segment has a less negative slope than the previous segment, the strain at the break-point between these segments can be eliminated from consideration. As shown for the situation of three strains, in this situation no cross involving this strain will have a higher value than a cross of two alternative strains. Thus it is possible to draw a "concave" broken line connecting those strains that should be considered as potential parental strains from the strain with the best production to the strain with the highest fertility. This line will be termed the "single line profit front", and is marked "SL-front" on Figure 15.3. All other strains will be below and to the left of this line.

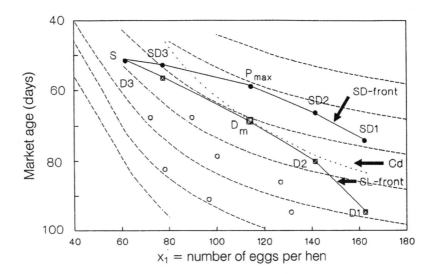

Figure 15.3. The most profitable cross among many strains of broilers. Broken lines are profit contours. S is the profit value of the sire strain. D_1, D_2, and D_3 are the profit values of the two dam lines. SD_1, SD_2 and SD_3 are the profit values of the crosses between S and the dam lines. D_m is the female parent of the optimum cross. P_{max} is the profit value of the optimum cross. Cd, the dam contour, is denoted with a dotted line.

Again starting with the strain with the best production, it is possible to perform a series of crosses between adjacent strains on the single line profit front. Each of these crosses will have the fertility value of the dam strain and the production value of the midpoint of the two parental strains. Thus a second broken line can be drawn above the single line profit front to represent the profit value of all possible crosses between strains on the single line profit front. This line will be termed the "sire-dam additive profit front", and is marked "SD-front" on Figure 15.3. These two fronts will meet only at their upper-left end, in which case the strain with the best production is used as the single parental strain.

As in the previous section, the best parental combination can be found by equating the tangent of each segment of the sire-dam additive profit front to the profit contours. Since the profit contours are convex, while the sire-dam additive profit front is concave, there will be only one point of tangency between them. Thus the cross of maximum profit will be found in the single segment in which the point of tangency between the profit contours and the sire-dam additive profit front is within the boundaries of the segment. Alternatively we note that for any given sire, $K_1 - K_2 y_s/2$ is a constant. Thus the potential dam lines can be ranked by the following formula:

$$C_D = K_2 y_d/2 + K_3/x_d \qquad [15.9]$$

Where C_D is the contribution of the dam to the cost of production. This equation can be rearranged so that y_d is a function of C_D as follows:

$$y_d = 2C_D/K_2 - 2K_3/(K_2 x_d) \qquad [15.10]$$

It is now possible on the profit map to plot the "dam contour" for any given dam line. Thus the optimum cross can also be found by equating the tangent of the dam contour with the single line profit front. The dam contour for the optimum cross is marked C_D on Figure 15.3. The optimum dam strain and the profit value of the cross are marked D_m and P_{max}, respectively. Due to the assumed additivity for the production trait, the slope of the dam contours is twice the slope of the profit contours. Similarly the slopes of the segments of the single line profit front will be twice the slope of the corresponding segments of the sire-dam additive profit front.

15.5 Choice of the most profitable parental combination when component traits are not genetically additive

In the previous sections genetic additivity on the scale of measurement was assumed for both traits. Thus of the five types of heterosis defined in the previous chapter only nonlinearity heterosis and sire-dam heterosis were considered. Nonlinearity heterosis was included because the scale of genetic additivity was nonlinear to the scale of profit. Therefore the profit contours were curved on the x and y scales. We will now show how the three additional types of heterosis, particularly heterosis of component traits and maternal effects, can be included in determination of the economically optimum cross.

In the example considered so far of a broiler production enterprise, profit will be a function of the broiler's and her dam's phenotypes. The only trait of economic importance in the sire's phenotype will be male fertility, and its economic value will be negligible relative to the former two traits. To utilize all possible forms of heterosis would require that the sire, the dam, and the offspring all be crossbreeds. Since the phenotype of the sire is of minimal importance, we will consider only heterotic effects on the offspring and the dam. In any cross between two lines there can be both a maternal effect and heterosis-of-component-traits effect on the productivity of the offspring. Although we assumed previously that the only trait of importance in the offspring is productivity, heterosis can also affect the viability of the offspring. If the dam is a crossbreed then there can be both types of heterotic effects on her reproductive rate. Even if only two lines are available, it is still possible to obtain heterosis for the dam's fertility by backcrossing the F-1 to one of the parental strains.

On a profit map the effect of heterosis will be to move the sire-dam profit front farther to the right. However, unlike the previous situation of assumed additivity, the amount of heterosis that will be obtained from any particular cross cannot be predicted accurately. For example if two lines are crossed to produce an F-1 that exhibits heterosis for production, and these progeny are then backcrossed to the dam strain to obtain heterosis for female reproduction, it is difficult to predict how much of the heterosis for production will be retained by the backcross progeny, which are now the commercial broilers. Even if it is assumed that half of the heterosis will be retained, this will not have an equal effect on profit, since the profit contours are curved relative to the scales of trait measurement.

If there is heterosis for component traits, then several of the conclusions of the previous section no longer hold. In the previous section we demonstrated that strains below and to the left of the single line profit front can be excluded from consideration as potential parents. However, due to heterosis of component traits, it is possible that a cross between lines below the single line profit front may be the most profitable. Also, due to heterosis, two lines with equal profitability may result in different trait values, and therefore different profitability for a cross. In the previous section we explained how even in the absence of heterosis for component traits, the most profitable dam line may in fact be a cross between two lines. In the previous sections we noted that, under the assumption of genetic additivity, there is no need to maintain the two parental dam lines separately. However, if there is heterosis for reproductivity, then dams produced as an F-1 cross between the two maternal parental strains may be more profitable than a "synthetic" dam strain that would result from repeated crossing among later generations of the original cross.

In conclusion, with heterosis, the most profitable cross can only be found by field trials and comparison among the different possibilities. If more than two or three strains are available, the number of possible crosses becomes quite large, especially if all possible three-way and four-way crosses are included. Although we can no longer use the absolute rules developed in the last section to exclude strains from consideration, it does seem reasonable to adopt several criteria to limit the number of possibilities. First, the strain with the better reproductivity should be used as dam strains, or parents of dam strains. Second, strains clearly inferior for all traits of interest can be excluded from consideration. Third, it is unlikely that a four-way or five-way cross can be significantly better than the best three-way cross. Finally, maintaining multiple lines is expensive. Thus, a more complicated crossing structure should be accepted only if a significant increase in profit can be demonstrated over the best simpler crossing structure.

15.6 Summary

Using the graphic method of Moav, strategies were developed to determine the most profitable cross among a number of potential parental strains. Profit was assumed to be a linear function of a single production trait, and an inverse function of a single reproduction trait. The method of equating tangents was used to determine the optimum cross under the assumption of genetic additivity for both traits on the scale of measurement. If there is heterosis for component traits and maternal effects, the best cross can only be determined by field trials. Furthermore, if the profit equation is nonlinear, the same degree of heterosis for component traits can have a different effect on profit, depending on the trait values. Since both the dam and her offspring directly contribute to profit, efficiency of production may be increased if both are crossbreeds. In this case it will be necessary to maintain at least three separate parental strains.

Chapter Sixteen

Planned Matings Together with Line Breeding

16.1 Introduction

Until this point in our evaluation of heterosis we ignored selection within the parental lines, and assumed complete heritability of the traits affecting the profitability of a cross. In this chapter we will develop methods to economically evaluate crossbreeding together with selection within the parental lines under the assumption of incomplete heritability for the component traits. Most of this chapter will be based on Moav and Hill (1966), and we will again consider primarily the case of a single production and a single reproduction trait. Long-term selection for both traits in a single line will be compared to selection within separate sire and dam lines. Comparison of alternative breeding schemes will be based on the graphic method of Moav and Hill (1966) which we explained in Section 9.8. We will show how the tangents of the profit contours and the response ellipse can be equated to determine the optimum selection index for parental lines used to produce crossbred progeny.

16.2 Derivation of the optimum selection index for a single line

In Chapter 9 we discussed the various methods that have been suggested to determine selection indices for situations in which the economic criterion is not a linear function of the trait values. For the linear situation, selection of the individuals with the highest estimated aggregate genotype will result in the greatest economic response to selection. However, as shown by Goddard (1983) this is not the case for nonlinear profit functions. Thus Wilton, Evans, and Van Vleck (1968) showed that for the case of a quadratic profit function, a nonlinear selection index will maximize the expected mean breeding value of the selected individuals. However, maximum long-term genetic progress will always be obtained by a linear selection index.

In Sections 9.8 and 9.9 we showed how the optimum linear index, i.e. the index that will result in maximum increase in the economic criteria, can be derived by equating the tangents of the "response ellipse" and the profit contours.

If the profit contours are curved then the optimum index will be a function both of the original trait values and the "size" of the response ellipse. That is the optimum index will be different for a low vs. a high selection intensity. Also, as shown by Goddard (1983) the optimum selection index to achieve maximum gain in a single generation will be different from the optimum selection index to achieve maximum gain over several generations.

In Chapters 6, 7, 9, 14 and 15 we considered the example of a broiler or swine production enterprise. We assumed that all individuals were selected on the same index, and that profit was a linear function of a production trait, such as growth rate, and an inverse function of a reproductive trait, for example the number of viable progeny per dam. Profit per broiler, which we will denote simply as P, was computed using the following equation, which appeared previously as equations [6.21], [7.9], [9.75], [14.4] and [15.1]:

$$P = K_1 - K_2 y - K_3/x \qquad [16.1]$$

Where y is the production trait value, for example age to reach market weight; x is the reproduction trait, for example number of live progeny per dam; and K_1, K_2 and K_3 are economic constants. For simplicity, x_1 and x_2 of the equations [6.21], [7.9], [9.75], and [14.9] were replaced with x and y. Assuming a linear index, I_o of the form given in equation [9.49], the index coefficients that result in the greatest economic progress, under the assumption that x and y are uncorrelated were derived in equation [9.76]. We will repeat these two equations here:

$$I = b_x x + b_y y \qquad [16.2]$$

$$\frac{b_y}{b_x} = \frac{-x^2 K_2 h_y^2}{K_3 h_x^2} \qquad [16.3]$$

Where h_x^2 and h_y^2 are the heritabilities of x and y, respectively. As is generally true for selection index, it is only the ratio of the coefficients which is important, not their absolute values. We note that this ratio is a function of x^2. Thus as reproductivity increases, the economic value of this trait in the selection index decreases, and emphasis in selection will shift to the production trait. The correlated responses of the component traits, ϕ_x and ϕ_y, were computed in equations [9.52] and [9.53] as follows:

$$\phi_x = i b_x (h_x \sigma_x)^2 / \sigma_I \qquad [16.4]$$

$$\phi_y = i b_y (h_y \sigma_y)^2 / \sigma_I \qquad [16.5]$$

Where i is the selection intensity, and σ_x, σ_y, and σ_{I_o} are the standard deviations of x, y and the selection index. The variance of the selection index was derived in equation [9.50] as follows:

$$(\sigma_{Io})^2 = (b_x\sigma_x)^2 + (b_y\sigma_y)^2 \qquad [16.6]$$

Substituting the values for b_x, b_y, and σ_{Io} from equations [16.3] and [16.6] into equations [16.4] and [16.5] gives the following formula for ϕ_x and ϕ_y:

$$\phi_x = \frac{iA_x}{K_3\sqrt{A_x + x^4A_y}} \qquad [16.7]$$

$$\phi_y = \frac{-iA_yx^2}{K_2\sqrt{A_x + x^4A_y}} \qquad [16.8]$$

Where $A_x = K_3^2h_x^4\sigma_x^2$ and $A_y = K_2^2h_y^4\sigma_y^2$. For a given selection intensity, selection on the index of equations [16.2] and [16.3] will maximize the mean aggregate genotype of the offspring in the following generation. However, for a situation of differing sire and dam contributions to profit, this is not equivalent to the index that will result in maximum profit in the next generation.

16.3 Derivation of economically optimum sire and dam indices for maximum profit in the next generation

In the previous two chapters we have dealt primarily with a case in which the dam and sire have different contributions to the total profitability of the cross. We will now develop formulas for the optimum selection index for the case considered in detail in the previous two chapters. Since the offspring are generally sold for slaughter before they reach the age of reproduction, the economic value of the reproduction trait will be a function of the dam's phenotype, while the economic value of the production trait will be a function of the offspring's phenotype. Assuming genetic additivity and complete heritability, the offsprings's phenotype will be equal to the mean of the parental values. Using these assumptions we derived equation [14.9], which we will repeat here with minor modifications:

$$P_{sd} = K_1 - K_2(y_s + y_d)/2 - K_3/x_d \qquad [16.9]$$

Where P_{sd} is the profitability of the particular cross, y_s and y_d are the dam and sire values for the production trait, and x_d is the dam value for the reproduction trait. Again we replaced x_1 and x_2 of equation [14.9] with x and y. The contribution of the sire to the cost of production is $K_2 y_s / 2$. In equation [15.9] we showed that the contribution of the dam to the cost of production, C_D can be derived from equation [16.9] as follows:

$$C_D = K_2 y_d / 2 + K_3 / x_d \qquad [16.10]$$

The optimum dam index can be derived by equating the tangent of the response ellipse to the tangent of C_D. The derivatives of the sire contribution will be zero and $K_2 / 2$ for x and y, respectively. Thus the optimum sire index will be direct selection on y. The optimum dam index can be derived by the method described in sections 9.8 and 9.9. The derivatives of C_D with respect to x and y are $-K_3 / x_d^2$ and $K_2 / 2$, respectively. These derivatives are equal to a_x and a_y of equation [9.59]. Substituting these values into equation [9.59] gives:

$$\frac{b_y}{b_x} = \frac{-x_d^2 K_2 h_y^2}{2 K_3 h_x^2} \qquad [16.11]$$

Which is the same as equation [16.3], except for a factor of 2 in the denominator. Thus for the optimum dam index, twice as much emphasis will be placed on the reproductivity trait for a given value of x.

Selection of sires and dams on these indices will result in the most profitable cross in the next generation, but will not result in the greatest possible gain in the aggregate genotype of the progeny. This is illustrated in Figure 16.1 for a swine enterprise, using values adapted from Moav and Hill (1966) for the contemporary situation in the United Kingdom. Profit is measured in pence per progeny. In this example $K_2 = 6.8$ pence/day, $K_3 = 13,000$ pence/progeny, $i = 1$, $h_x^2 = 0.1$, $\sigma_x = 5$ progeny, $h_y^2 = 0.4$, and $\sigma_y = 25$ days. The initial values for x and y are 10 progeny and 200 days. The points S and D_m denote the genetic gain in one generation, with sires selected only on growth rate, and dams selected by the linear index with the index coefficients of equation [16.11]. The point I denotes the maximum genetic gain if all individuals are selected on the optimum single line index, using the index coefficients of equation [16.3]. All three points are on the response circle centered at O. The point O' is the mean breeding value of the cross between sires and dams selected on separate indices. Note that O' is within the response circle and on a lower profit contour than I. The point SD_m denotes the profit value of the cross between S and D_m. Note that this point is outside the response circle, and on a higher profit contour than S, D_m, I, or O'. Thus, selection on separate sire and dam indices can result in a greater mean profit in the next generation than that obtained by selection on the optimum single index.

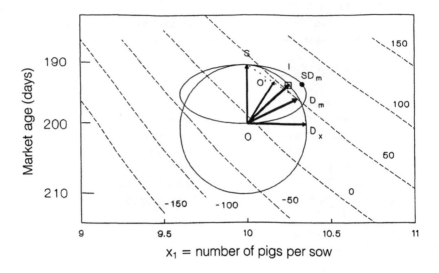

Figure 16.1. Selection of male and female pigs on specialized indices. Broken lines are profit contours. O is the trait values prior to selection. S and D_m are the genetic gain in one generation, with sires selected only on growth rate, and dams selected by the linear index with the index coefficients of equation [16.11]. The point I denotes the maximum genetic gain if all individuals are selected on the optimum single line index, using the index coefficients of equation [16.3]. O' is the mean breeding value of the cross between sires and dams selected on separate indices. SD_m is the profit value of the cross between S and D_m.

If the sires are selected only on growth rate, while the dams are selected on an index of both traits, the profit values of all possible crosses will describe an ellipse denoted the "sire-dam profit ellipse". For sires the response to selection, ϕ_{ys}, will be $i_s h_y^2 \sigma_y$, where i_s is the selection intensity for sires. Following equation [9.60] the response in standardized units, ϕ_{ys}^*, is computed as follows:

$$\phi_{ys}^* = dy/(h_y^2 \sigma_y) = i_s \qquad\qquad [16.12]$$

The response of the dams for y in standardized units can be derived from equation [9.61] as follows:

$$(\phi_{xd}^*)^2 + (\phi_{yd}^*)^2 = i_d^2 \qquad\qquad [16.13]$$

$$\phi_{yd}^* = [i_d^2 - (\phi_{xd}^*)^2]^{1/2} \qquad\qquad [16.14]$$

Where i_d is the selection intensity for dams. The profit values for the progeny in standardized units will be:

$$\phi_{ysd}^{\ *} = (\phi_{ys}^{\ *} + \phi_{yd}^{\ *})/2 \qquad\qquad [16.15]$$

$$\phi_{xsd}^{\ *} = \phi_{xd}^{\ *} \qquad\qquad [16.16]$$

where $\phi_{ysd}^{\ *}$ and $\phi_{xsd}^{\ *}$ are the profit values of the cross for y and x, respectively. (These values are *not* the correlated responses to selection of the progeny, which will be equal to the parental means for both traits.) Substituting equations [16.12] and [16.14] into equation [16.15] gives:

$$\phi_{ysd}^{\ *} = i_s/2 + \{[i_d^{\ 2} - (\phi_{xsd}^{\ *})^2]^{1/2}\}/2 \qquad\qquad [16.17]$$

On rearrangement, the following ellipse is obtained:

$$\frac{(\phi_{xsd}^{\ *})^2}{i_d^{\ 2}} + \frac{(\phi_{ysd}^{\ *} - i_s/2)^2}{(i_d/2)^2} = 1 \qquad\qquad [16.18]$$

The sire-dam response ellipse is also illustrated on Figure 16.1. The center of the ellipse relative to the base population will be at the point $(0, i_s/2)$. The axes will be parallel to the co-ordinate axes, and of length $2i_d$ and i_s, for x and y, respectively. Note that a considerable section of the sire-dam response ellipse is outside the response circle to the right. Any of these crosses will result in greater profit than can be obtained by selection of all individuals on a single index.

16.4 Heterosis and line breeding, long-term considerations

Although separate sire and dam indices will result in the greatest gain in profit in the next generation, it will not necessarily result in the greatest long-term gain in profit. In the example given above, the profit value of the cross will be at SD_m. However, the mean genetic level of the progeny will be at O', and this will be the base level of the population for selection in the next generation. Thus the strategy described in the previous section does *not* maximize genetic progress or mean profit value of the offspring over several generations. Moav and Hill (1966) compared three long-term breeding strategies:

1. Selection of both males and females on the optimum single-line index.
2. Splitting of the original population into separate sire and dam lines each selected on the economically optimal index for that sex.

3. Selection of males and females on individually economically optimal indices within a single line.

In evaluating these strategies we are first confronted with the problem that the profit contours are nonlinear. Thus the index for maximum gain in one generation will not be the same as the index for maximum gain in several generations. However, gains made in later generations should be discounted more heavily than gains made in earlier generations. For simplicity we will assume that the selection intensity is equal to unity in all cases, and that the goal is to maximize progress in four generations.

Using the values given for Figure 16.1, we will first compute the optimum index for selection in a single line. From equation [16.3]:

$$\frac{b_y}{b_x} = \frac{-x^2 K_2 h_y^2}{K_3 h_x^2} = -x^2(0.002092) = -(x_o + \phi_{xm})^2(0.002092) \qquad [16.19]$$

Where x_o = 10 is the original number of progeny per dam, and ϕ_{xm} is the response of x to selection on the optimum linear index. ϕ_{xm} can be derived from equation [16.7] as follows:

$$\phi_{xm} = \frac{iA_x}{K_3\sqrt{A_x + x^4 A_y}} = \frac{13,000}{\sqrt{42,250,000 + 4624(10 + \phi_{xm})^4}} \qquad [16.20]$$

Although the selection intensity is equal to unity in each generation, the objective is to determine the index that results in maximum progress after four generations. Therefore the value of i = 4 is used in equation [16.20]. This equation has only one unknown, ϕ_x, but it is a complex function of this value. Solving by iteration, ϕ_x = 1.21, and the value for x after selection is 11.21 offspring. Inserting this value in equation [16.19] gives b_y/b_x = $-(11.21)^2(0.002092)$ = -0.262. The correlated response in y, ϕ_{ym} can be computed from equation [16.8] as follows:

$$\phi_{ym} = \frac{-iA_y x^2}{K_2\sqrt{A_x + x^4 A_y}} = \frac{-2720(11.21)^2}{\sqrt{42,250,000 + 4624(11.21)^4}} =$$

$$= -31.83 \text{ days} \qquad [16.21]$$

Thus the value for y after selection is 200 − 31.83 = 168.17 days. Under the assumption that pigs are slaughtered at a constant weight, genetic improvement increases profit by decreasing the variable costs of production, which can be computed from equation [16.1] as: $K_2 y + K_3/x$. The original value was 6.8(200)

+ 13,000/10 = 2680 pence. The new value for the variable cost can then be computed as follows:

$$K_2 y' + K_3/x' = 6.8(168.17) + 13,000/11.21 = 2303.2 \text{ pence} \qquad [16.22]$$

Where x' and y' are the trait values after selection.

We will now compute the optimum sire and dam indices if the population is split into separate sire and dam lines. If the sire line is selected only on y, then the gain in one generation will be −0.4(25) = −10 days, or 40 days in four generations. (The maximum gain that could be made for selection only on reproduction rate would be (0.1)(5) = 0.5 offspring in one generation, or 2 offspring in four generations.) The optimum dam index can be found by substituting $2K_3$ for K_3 in equations [16.19], [16.20], and [16.21]. The responses of x and y to selection on this index, ϕ_{xd} and ϕ_{yd}, are computed as follows:

$$\phi_{xd} = \frac{2iA_x}{K_3\sqrt{4A_x + x^4A_y}} = \frac{26,000}{\sqrt{169,000,000 + 4624(10 + \phi_{xd})^4}} =$$

$$= 1.63 \text{ progeny} \qquad [16.23]$$

$$\phi_{yd} = \frac{-iA_y x^2}{K_2\sqrt{4A_x + x^4A_y}} = \frac{-2720(11.63)^2}{\sqrt{169,000,000 + 4624(11.63)^4}} =$$

$$= -23.10 \text{ days} \qquad [16.24]$$

As in the first breeding scheme, these equations are evaluated with i=4. The value for the variable costs after selection of the sire and dam line cross can then be evaluated by inserting these values into the expression for the variable costs in equation [16.9] as follows:

$$K_2(y_s + y_d)/2 + K_3/x_d = 6.8(160 + 176.90)/2 + 13,000/11.63 =$$

$$= 2263.3 \text{ pence} \qquad [16.25]$$

Thus the second alternative results in a reduction of 40 pence per pig sold after four generations, or a 2% increase in genetic gain for profitability.

For the third breeding scheme, it is necessary to first estimate the genetic gain for a single generation of selection with separate sire and dam indices. In this case, i = 1. Thus, for the sires the gain will be $h_y^2\sigma_y = -0.4(25) = -10$ days. The correlated responses for selection on the optimum dam index can then be derived from equations [16.23] and [16.24] with i = 1, as follows:

$$\phi_{xd} = \frac{2iA_x}{K_3\sqrt{4A_x + x^4A_y}} = \frac{6500}{\sqrt{169,000,000 + 4624(10 + \phi_{xd})^4}} =$$

$$= 0.43 \text{ progeny} \qquad\qquad [16.26]$$

$$\phi_{yd} = \frac{-iA_yx^2}{K_2\sqrt{4A_x + x^4A_y}} = \frac{-680(10.43)^2}{\sqrt{169,000,000 + 4624(10.43)^4}} =$$

$$= -4.94 \text{ days} \qquad\qquad [16.27]$$

The trait responses for a single generation are nearly equal to one fourth of the response for four generations, but the optimum index for one generation gives slightly greater emphasis to reproduction. If the population is not split into two separate breeding populations, then the mean trait values will be the means of the progeny of the two selected groups. Thus:

$$\phi_x = (x_s + x_d)/2 - x_o = (10 + 10.43)/2 - 10 = 0.215 \text{ progeny} \qquad [16.28]$$

$$\phi_y = (y_s + y_d)/2 - y_o = (190 + 195.06)/2 - 200 = -7.47 \text{ days} \qquad [16.29]$$

In three generations the gains will be approximately three-fold these values, that is 0.645 progeny, and 22.41 days. In the final generation there will be an additional gain of -10 days for y in the sires and 0.43 and -4.94 for x and y for dams. Thus the trait values will be $x_s = 10.654$ and $y_s = 167.59$ for sires, and $x_d = 11.075$ and $y_d = 172.65$. The reduction in the variable costs in the fourth generation progeny will then be:

$$K_2(y_s + y_d)/2 + K_3/x_d = 6.8(167.59 + 172.65)/2 + 13,000/11.075 =$$

$$= 2330.6 \text{ pence} \qquad\qquad [16.30]$$

This alternative results in the lowest long-term gain, although even this alternative is only 67 pence less than the best alternative (3%), and only 27 pence less than the first alternative. It should be noted though, that this alternative results in the maximum gain in profit in the first generation of selection.

We should note here again a point that we have mentioned previously several times. Selection index is highly robust. Relatively large changes in the economic values, and therefore the index coefficients of the component traits, have only minor effects on the correlated responses of the individual traits. Thus it is not surprising that all three alternatives resulted in very similar responses. Thus although maximum progress was obtained by splitting the original population into separate sire and dam lines, it is hard to justify this scheme, except for species with short generation intervals and high fecundity.

16.5 Choice of the optimum lines for line breeding together with cross breeding

In the previous chapter we considered the question of the optimum cross when several lines are available, but without considering selection within the parental lines. We concluded that the line with the highest production should be used as the sire line. Assuming that maternal effects and heterosis of component traits are insignificant, then lines on the same dam contour, i.e., equal C_D values, will result in crosses of equal profitability. If selection is practiced within the sire and dam lines, the long-term profitability of dam lines with equal C_D values can differ. This will be illustrated by continuing the example of the previous section.

C_D of the line of swine considered in the previous section can be computed from equation [16.10] as follows:

$$C_D = K_2 y_d/2 + K_3/x_d = 6.8(200)/2 + 13000/10 = 1980 \qquad [16.31]$$

Assume an additional line, D_1, with equal C_D, but with $x_{d1} = 14$. In this case $y_{d1} = 309.2$ days. Although these lines have equal C_D the profitability of the new line will be lower. The variable cost of production, $K_2 y_d + K_3/x_d$, is 2660 pence for the original line, and 3031 pence for the new line. This is because the slopes of the dam contours are twice the slopes of the profit contours. In the previous section we assumed that a single line was used to breed both sires and dams on individual indices. The variable costs of production of a cross produced from one generation of selection of sires and dams from this line can be computed from equation [16.25] as follows:

$$K_2(y_s + y_d)/2 + K_3/x_d = 6.8(190 + 195.06)/2 + 13,000/10.43 =$$

$$= 2555.6 \text{ pence} \qquad [16.32]$$

We will now assume that the line considered in the previous section is still used as the sire line, but D_1 is used as the dam line. The optimum sire index will still be selection on y. The responses to selection on the optimum dam index, ϕ_{xd1} and ϕ_{yd1}, can be computed from equations [16.26] and [16.27] with the new values for this line.

$$\phi_{xd1} = \frac{2iA_x}{K_3\sqrt{4A_x + x^4 A_y}} = \frac{6500}{\sqrt{169,000,000 + 4624(14 + \phi_{xd1})^4}} =$$

$$= 0.34 \text{ progeny} \tag{16.33}$$

$$\phi_{\text{ydl}} = \frac{-iA_y x^2}{K_2\sqrt{4A_x + x^4 A_y}} = \frac{-680(14.34)^2}{\sqrt{169,000,000 + 4624(14.34)^4}} =$$

$$= -7.32 \text{ days} \tag{16.34}$$

Note that for this line the correlated response is less for x, but greater for y. It should now be clear that since there is an inverse relationship between profit and reproduction, the reproductive trait should be given less emphasis in selection for a strain with a higher reproductive rate. Equation [16.25] can again be used to compute the economic value of this cross, as follows:

$$K_2(y_s + y_{\text{dl}})/2 + K_3/x_{\text{dl}} = 6.8(190 + 192.68)/2 + 13,000/10.34 =$$

$$= 2558.4 \text{ pence} \tag{16.35}$$

Thus profit for this cross is 5 pence less than profit from sire and dam selection from the original line. This is again due to the inverse relationship between profit and reproduction. Since D_1 is already at a higher reproductive level, the additional gain in reproduction has less of an effect on profit than for the original strain. Of course a line with the same reproductive level as D_1, but with slightly higher production although still less than the original line would result in a more profitable cross. A line with these trait values would still be situated on a lower profit contour than the original line.

From this example it is apparent that the choice of the optimal parental lines is even more complicated once selection within each line is considered. It will not only be a function of the trait values, but also a function of the discount rate and the profit horizon. More exact calculations have not been worked out, to the best of my knowledge.

16.6 Summary

In this final chapter we extended the graphic method of Chapter 9 to determine the optimum selection index and the correlated trait responses for the case of a production and a reproduction trait. Profit was assumed to be a linear function of production and an inverse function of reproduction. We demonstrated that the profit of a cross between sires and dams selected on different indices can have a higher profit value than any point on the response ellipse.

Three breeding schemes were compared: 1) selection of all individuals on the optimum selection index, 2) splitting of the original population into separate ы1ь aлd dam lines each selected on the economically optimal index for that sex, and 3) selection of males and females on separate indices within a single line. Scheme 2 gave maximum long-term increase in profit, and scheme 3 gave the least. For a single generation, maximum profit was realized by scheme 3, followed by scheme 2.

If selection is practiced within parental lines that are then crossed to produce the commercial offspring, then dam lines on the same dam contour can result in differing profit values after selection.

GLOSSARY OF SYMBOLS

Matrices and vectors are listed in **bold** type. Matrices and Vectors are listed before scalars with the same symbol. Symbols with capital letters are listed before symbols with the same lower case letters. The section in which the symbol is first mentioned is listed in parenthesis after the definition. Greek symbols are given after the Latin symbols.

Latin Symbols

A   Matrix of quadratic economic values (8.5)

a   Vector of economic values (3.4)

A_l Price of unit product (6.2)

AC   Average costs (2.8)

A_c Accuracy of genetic evaluation (13.3)

A_d Price of unit of dam product (5.2)

A_o Price of unit of offspring product (5.3)

b_{op} Regression of offspring on parent (1.5)

B   Matrix of quadratic index coefficients (8.4)

B   Annual income per animal selected and per unit of selection intensity (11.4)

b   Vector of selection index coefficients (3.4)

BV Breeding value (1.6)

C Genetic covariance matrix among traits in x and y (3.4)

C Costs (4.5)

C_c Annual costs of a breeding program (8.3)

C_D Contribution of the dam to the costs of production (15.4)

C_d Unit feed costs (5.4)

C_i Initial costs of a breeding program (11.4)

C_n Non-feed costs per enterprise (7.2)

C_W Investment (7.2)

D Matrix that describes the passage of genes from time t-1 to time t (8.4)

D Days from weaning to slaughter (5.4)

d Vector of discounting factors (8.4)

D_c Net present value discounting factor for annual costs of a breeding program

(11.5)

d_i "Nominal" interest rate (2.7)

d_{it} Vector of discounting factors for consecutive years (8.4)

d_k Risk on investment (8.3)

D_{max} Optimum slaughter age for maximum economic efficiency (5.5)

d_o Opportunity cost discount rate (8.2)

d_p Change in profit due to index selection (9.8)

d_q "Real" discount rate, corrected for inflation (2.9)

D_r Net present value discounting factor for annual returns of a breeding program (11.5)

d_r Required "nominal" interest corrected for inflation, tax, and risk (8.2)

d_s The social time preference discount rate (8.2)

d_t Inflation rate (2.9)

DV   Discounted value (2.9)

d_x Tax rate on investment (8.2)

E   Economic efficiency (4.5)

e   Environmental effect or residual (1.3)

E_B Biological efficiency (7.2)

E_d Elasticity of demand (2.4)

E(f) Expectation of f (9.3)

EF_d Feed costs per dam (5.1)

EF_o Feed costs per offspring (5.1)

E_i Inverse of economic efficiency (7.3)

E_s Elasticity of supply (2.4)

F   Quantity of feed given per enterprise (7.2)

F_1 Fixed costs per unit product (6.2)

F_2 Fixed costs per animal (6.2)

F_d Annual feed costs of the enterprise (5.4)

F_{Md} Maintenance feed per unit metabolic body weight of the dam (5.4)

F_{Mo} Maintenance feed required per unit weight of offspring per day (5.4)

F_{Pd} Feed required by the dam per offspring produced (5.4)

F_{Po} Feed required per unit product of offspring (5.4)

G Variance matrix for traits in y (3.4)

g Genetic effect (1.3)

G_f Genetic superiority of females selected as parents for the next generation (8.4)

G_m Genetic superiority of males selected as parents for the next generation (8.4)

H Aggregate genotype (3.4)

h^2 Heritability (1.5)

H_{nl} Non-linearity heterosis (14.4)

H_q Aggregate quadratic genotype (9.2)

H_{sd} Sire-dam heterosis (14.4)

I A linear selection index (9.8)

i Selection intensity (1.7)

I_d Non-feed costs of dam (5.1)

I_l Optimum linear selection index (9.3)

I_m Optimum sire-dam mating index (14.5)

I_o Non-feed costs of offspring (5.1)

I_q Optimum quadratic selection index (9.3)

I_s Selection index (3.4)

I_W Return on investment (7.2)

J Matrix so that $\mathbf{J'J} = \mathbf{P}$ (9.8)

K Income less fixed costs per unit product (6.2)

K_A Costs of breeding program per animal examined (11.4)

K_F Fixed costs of breeding program (11.4)

K_s Fixed costs per daughter of sire sampled (13.3)

K_1 Income per gram egg less fixed costs per gram egg (6.5)

K_2 Fixed costs per egg (6.5)

K_3 Fixed costs per unit weight of animal (6.3)

K_4 Fixed costs per animal (6.3)

L   Generation interval (1.7)

M   Matrix to account for gene flow over generations and time (8.5)

M   Number of selected individuals (11.4)

m   Vector of the value of trait expressions in an animal in consecutive years

(8.4)

m   Number of animals per enterprise (6.4)

m_0 Original number of animals per enterprise (6.4)

MC Marginal costs (2.8)

m_d Number of dams per enterprise (5.2)

m_t Vector of proportion of genes at time t in all age-sex classes (8.4)

MU Marginal utility (2.8)

N   Current value in monetary units (2.9)

N_t Discounted gain from selection to time t (8.4)

n_d Number of semen doses per sire (13.3)

n_s Number of daughters sampled per sire (13.3)

O   Production costs as a function of the offsprings's genotype (14.2)

P   Phenotypic variance matrix for traits in **x** (3.4)

P   Profit (4.5)

p   proportion selected (1.7)

P_1 Profit per unit product (6.2)

P_2 Profit per animal (6.2)

P_3 Profit per unit weight of animal (6.3)

PD Predicted difference (1.6)

P_d Profit of a dam line (14.4)

P_E Profit per enterprise (6.4)

P_M Profit for fixed number of animals (6.4)

P_m Mean of the profit values of two parental lines (14.4)

p_{max} The proportion selected for which profit is maximum (11.4)

P_o Profit value of a cross between two lines (14.4)

P_Q Profit with fixed demand (6.4)

P_r Price (2.4)

P_s Profit of a sire line (14.4)

P_{sd} Profit of a cross between sire line s and dam line d (16.3)

P_W Profit with fixed investment (6.4)

Q   Matrix that includes changes in the population structure due to ageing (8.4)

Q   Quantity of production (2.4)

R   Revenue or returns (2.4)

r   Discounting factor (8.3)

R_d Returns from female production (5.1)

r_g Genetic correlation (3.3)

r_{HI} Correlation between the selection index and the aggregate genotype (3.5)

R_M Number of individuals examined for each individual selected (13.3)

R_o Returns from offspring production (5.1)

rpt   Repeatability (1.5)

RSE Relative selection efficiency of an alternate index vs. the optimum index

(3.5)

R_t Total returns from continuous selection to time t (8.4)

r_t Returns from one generation of selection at time t (8.3)

S   Mean value of selected individuals in trait units (1.7)

T   Years to profit horizon (8.3)

t   Time in years (2.9)

T_{max} Value for T at which profit is maximum (11.4)

T_n Number of individuals examined (11.4)

U   Matrix for probability of trait expression in progeny times fraction of

ancestor's genotype expressed for a given year-parity combination (8.4)

u   Vector of the fraction of an ancestor's genotype expressed in his progeny

in consecutive years (8.4)

V   Minimal acceptable annual return, also the value from one year of genetic

improvement (2.9)

v   Vector of economic values of gains in selection from generation 1 to t (8.4)

W   Total weight of animals per enterprise (6.4)

x   Vector of measured traits (3.4)

x   Phenotypic trait value (1.3)

x_D Quantity of product per dam (5.2)

x_d Value of reproductive trait for the dam line (15.2)

x_m Value of reproductive trait for the dam of the most profitable cross (15.2)

x_s Value of reproductive trait for the sire line (15.2)

x_1 Number of offsprings (eggs) marketed per dam per year (5.3)

x_{10} Original number of offsprings (eggs) per dam (6.4)

x_{1o} Number of offsprings marketed/female/year (5.3)

x_2 Mean weight of eggs (6.3)

x_{2o} Quantity of product per offspring (5.3)

x_{2d} Product per unit weight of dam (5.2)

x_3 Body weight of animal (6.3)

x_{3d} Mean weight of dams (5.2)

x_{30} Original weight of animals prior to genetic improvement (6.4)

x_{3o} Mean weight of progeny (5.5)

x_{4o} Daily gain of progeny (5.5)

y   Vector of breeding values for traits of economic importance (3.4)

y_d Value of the production trait for the dam line (15.2)

y_{fo} Vector of original breeding values of females (8.4)

y_m Value of the offspring of the most profitable cross for the production trait (15.2)

y_{mo} Vector of original breeding values of males (8.4)

Y_{op} Predicted trait value for progeny of a parent (1.6)

Y_p Trait value of parent (1.6)

y_s Value of the production trait for the sire line (15.2)

Z Matrix to describe passage of genes from one time unit to the next (8.4)

z ordinate of the normal curve (1.7)

Greek Symbols

α Scaling factor or proportionality constant (7.5)

δ Vector of proportional changes in economic traits (9.7)

δ Partial derivative (7.5)

μ Vector of trait means (9.2)

σ_A^2 Additive genetic variance (1.5)

σ_{axy} Genetic covariance between traits x and y (3.3)

σ_E^2 Environmental variance (1.5)

σ_g^2 Genetic variance (1.5)

σ_H^2 Variance of the aggregate genotype (3.4)

σ_{HI} Covariance between the aggregate genotype and the selection index (9.6)

σ_I^2 Variance of any linear selection index (9.6)

$\sigma_{I_l}^2$ Variance of optimum linear selection index (9.4)

$\sigma_{I_s}^2$ Variance of the optimum selection index (3.5)

$\sigma_{I_q}^2$ Variance of optimum quadratic selection index (9.4)

σ_{op} Parent-offspring covariance (1.5)

σ_p^2 Phenotypic variance (1.4)

ϕ Vector of correlated responses of individual traits to selection on an index

(3.5)

ϕ Response to selection in trait units (1.7)

ϕ_x Response of trait x due to index selection (9.8)

ϕ_x^* Standardized response of trait x due to index selection (9.8)

ϕ_y Response of trait y due to index selection (9.8)

ϕ_y^* Standardized response of trait y due to index selection (9.8)

$\phi_{y/x}$ Correlated response of trait y to direct selection on trait x (3.3)

Γ Vector of Lagrange multipliers (9.7)

Δ Small change in parameter value (7.5)

ΔF Increase in inbreeding per generation (1.8)

ΔG Response to selection per year (1.7)

REFERENCES

Allaire, F. R. 1977. Corrective mating methods in context of breeding theory. *J. Dairy Sci.* **60**, 1799-1806.

Bondioli, K. R., S. B. Ellis, J. H. Pryor, M. W. Williams, and M. M. Harpold. 1989. The use of male-specific chromosomal DNA fragments to determine the sex of bovine preimplantation embryos. *Theriogenology* **31**, 95-104.

Brascamp, E. W. 1973. Model calculations concerning economic optimization of AI breeding with cattle. *Z. Tierzuchtg. Zuchtgsbiol.* **90**, 1-15.

Brascamp, E. W. 1984. Selection indices with constraints. *Anim. Breed. Abs.* **52**, 645-654.

Brascamp, E. W., C. Smith and D. R. Guy. 1985. Derivation of economic weights from profit equations. *Anim Prod.* **40**, 175-180.

Cartwright, T. C. 1970. The use of system analysis in animal science with emphasis on animal breeding. *J. Anim. Sci.* **49**, 817-825.

Cunningham, E. P. 1969. The relative efficiencies of selection indexes. *Acta Agri. Scand.* **19**, 45-48.

Cunningham, E. P., Moen, R. A., and T. Gjedrem. 1970. Restriction of selection indexes. *Biometrics* **26**, 67-74.

Cunningham, E. P. and Joan Ryan. 1975. A note on the effect of the accounting period on the economic values of genetic improvement in cattle populations. *Anim. Prod.* **21**, 77-80.

Dalton, G. E. (Ed.) 1975. *Study of Agricultural Systems.* Applied Science Publishers Ltd. London, UK.

Dekkers, J. C. M. and G. E. Shook. 1990. Economic evaluation of alternative breeding programs for commercial artificial insemination firms. *J. Dairy Sci.* **73**, 1902-1919.

Dekkers, J. C. M. and G. E. Shook. 1990a. Genetic and economic evaluation of nucleus breeding schemes for commercial artificial insemination firms. *J. Dairy Sci.* **73**, 1920-1937.

Dickerson, G. E. 1970. Efficiency of animal production - molding the biological components. *J. Anim. Sci.* **30**, 849-859.

Dickerson, G. E. 1978. Animal size and efficiency: basic concepts. *Anim Prod.* **27**, 356-379.

Dickerson, G. E. 1982. Principles in establishing breeding objectives in livestock. *Proc. World Congr. on Sheep and Beef Cattle Breed.* **1**, 9-22.

Dror, J. (Ed.) 1988. The Israeli Dairy Council Yearbook 1987/88. The Israeli Dairy Council. Tel Aviv, Israel.

Essl, A. 1981. Index selection with proportional restriction: another viewpoint. *Z. Tierzuchtg. Zuchgsbiol.* **98**, 125-131.

Ezra, E. and J. I. Weller. 1989. Can the Israeli breeding program be improved? *Meshek Habakar Vehahalav* **219**, 13-18. (In Hebrew).

Falconer. D. S. 1964. *Introduction to Quantitative Genetics.* Oliver and Boyd. Edinburgh, UK.

Ferris, T. A. and B. W. Troyer. 1987. Break even costs for embryo transfer in a commercial dairy herd. *J. Dairy Sci.* **70**, 2394-2401.

Foote, R. H. and P. Miller. 1971. Sex Ratio at Birth-Prospects for Control. pp 1-9. In: *ASAS Symposium.* C. A. Kiddy and H. D. Hafs (Eds).

Gibson, J. P. 1987. The options and prospect for genetically altering milk composition in dairy cattle. *Anim. Breed. Abs.* **55**, 231-243.

Gibson, J. P. 1989. Altering milk composition through genetic selection. *J. Dairy Sci.* **72**, 2815-2825.

Gjedrem, T. 1972. A study on the definition of the aggregate genotype in a selection index. *Acta Agr. Scand.* **22**, 11-16.

Godard, M. E. 1983. Selection indices for non-linear profit functions. *Theor. Appl. Genetic.* **64**, 339-344.

Graybill, A. F. 1969. *Introduction to Matrices with Applications in Statistics.* Wadsworth Publishing Company, Inc. Belmont, CA.

Gruebele, J. W. 1988. Impact of supply management programs on regional dairying: West Coast perspective. *J Dairy Sci.* **71**, 2310-2314.

Hansen, L. B., A. E. Freeman, and P. J. Berger. 1983. Yield and fertility relationships in dairy cattle. *J. Dairy Sci.* **66**, 293-305.

Harville, D. A. 1975. Index selection with proportional constraints. *Biometrics* **31**, 223-225.

Hazel, L. N. 1943. The genetic basis for constructing selection indexes. *Genetics* **28**, 476-490.

Henderson, C. R. 1963. Selection index and expected genetic advance. pp. 141-163. In: *Statistical Genetics and Plant Breeding.* W. D. Hanson and H. F. Robinson (Eds.). Publ. No. 982. National Academy of Sciences - National Research Council, Washington, DC.

Henderson, C. R. 1973. Sire evaluation and genetic trends. pp 10-41. In: *Proc Anim. Breed. Genet. Symp. in Honor of Dr. D. L. Lush.* ASAS and ADSA, Champaign, IL.

Henderson, C. R. 1976. A simple method for computing the inverse of a numerator relationship matrix used in prediction of breeding values. *Biometrics* **32**, 69-83.

Henderson, C. R. 1984. *Applications of linear Models in Animal Breeding.* University of Guelph. Guleph, Canada.

Hermas, S. A., C. M. Young, and J. W. Rust. 1987. Genetic relationships and additive genetic variation of productive and reproductive traits in Guernsey dairy cattle. *J. Dairy Sci.* **70**, 1252-1257.

Hill, W. G. 1971. Investment appraisal for national breeding programmes. *Anim. Prod.* **13**, 37-50.

Hill, W. G. 1974. Prediction and evaluation of response to selection with overlapping generations. *Anim. Prod.* **18**, 117-139.

Hill, W. G. and R. Thompson. 1978. Probabilities of non-positive definite between group or genetic covariance matrices. *Biometrics* **34**, 429.

Hudson, G. F. S. and L. D. Van Vleck. 1983. Levels of inbreeding in AI cows and the effects of inbreeding on yield, stayability, and calving interval. pp. 156-180. In: *Genetics Research Report to Eastern AI Coop.* Cornell University, Ithaca, NY.

Itoh, Y. and Y. Yamada. 1988. Selection indices for desired relative genetic gains with inequality constraints. *Theor. Appl. Genet.* **75**, 731-735.

Jacoby, S. L. S., Kowalik, J. S., and Pizzo, J. T. 1972. *Iterative Methods for Nonlinear Optimization Problems.* Prentice-Hall, Inc. Englewood Cliffs, NJ.

James, J. W. 1978. Index selection for both current and future generation gains. *Anim. Prod.* **26**, 111-118.

James, J. W. 1982. Construction, uses and problems of multitrait selection indices. *Proc. 2nd World Congr. Genet. Appl. Livest. Prod.* **5**, 130-139.

Johannsen, W. 1903. *Uber Erblickeit in Populationen und in reinen Linien.* G. Fisher, Jena. (See reference to Peters, J. A.).

Johannsen, I. and J. M. Rendel. 1972. *Genetics and Animal Breeding.* Oliver and Boyd. Edinburgh, UK.

Kashi, Y., M. Soller, E. M. Hallerman, and J. S. Beckmann. 1986. Restriction fragment length polymorphisms in dairy cattle genetic improvement. *Proc 3rd World Congr. Genet. Appl. to Livest. Prod.* **12**, 57-63.

Kempthorne, O, and A. W. Nordskog. 1959. Restricted selection indices. *Biometrics* **15**, 10-19.

Kislev, Y. and U. Rabiner. 1979. Economic aspects of selection in the dairy herd in Israel. *Australian J. Agricul. Econom.* **23**, 128-146.

Lande, R. and R. Thompson. 1990. Efficiency of marker-assisted selection in the improvement of quantitative traits. *Genetics* **124**, 743-756.

Lee, K. L., A. E. Freeman, and L. P. Johnson. 1985. Estimation of genetic change in the registered Holstein cattle population. *J. Dairy Sci.* **68**, 2629-2638.

Leitch, H. W., E. B. Burnside, and B. W. McBride. 1990. Treatment of dairy cows with recombinant bovine somatotrophin, genetic and phenotypic aspects. *J. Dairy Sci.* **73**, 181-190.

Lin, C. Y. 1978. Index selection for genetic improvement of quantitative characters. *Theor. Appl. Genet.* **52**, 49-56.

Litt, M. and J. A. Luty. 1989. A hypervariable microsatellite revealed by in vitro amplification of a dinucleotide repeat within the cardiac muscle actin gene. *Am. J. Hum. Genet.* **44**, 397-401.

Maijala, K. 1976. General aspects in defining breeding goals in fram animals. *Acta. Agr. Scand.* **26**, 40-46.

McClintock, A. E. and E. P. Cunningham. 1974. Selection in dual purpose cattle populations: defining the breeding objective. *Anim Prod.* **18**, 237-247.

McGilliard, M. L. 1978. Net returns from using genetically superior sires. *J. Dairy Sci.* **61**, 250-254.

Meijering. A. 1984. Dystocia and still birth in cattle - a review of causes, relations and implications. *Livest. Prod. Sci.* **11**, 143-177.

Mendel, G. 1866. *Versuch uber Pflanzen-Hybriden* (Experiments in plant hybridization). *Proc. Brunn Natural History Society.* (See reference to Peters, J. A.).

Miller, P. D. 1988. Implementing technology for genetic improvement: industry's view. *J. Dairy Sci.* **71**, 1967-1971.

Moav, R. 1966. Specialized sire and dam lines. I. Economic evaluation of crossbreeds. *Anim Prod.* **8**, 193-202.

Moav, R. 1966a. Specialized sire and dam lines. II. The choice of the most profitable parental combination when component traits are genetically additive. *Anim Prod.* **8**, 203-211.

Moav, R. 1966b. Specialized sire and dam lines. III. The choice of the most profitable parental combination when component traits are genetically non-additive. *Anim Prod.* **8**, 365-374.

Moav, R. 1973. Economic evaluation of genetic differences. pp 319-352 in *Agricultural Genetics, Selected Topics.* R. Moav (Ed.) John Wiley and Sons. New York, NY.

Moav, R. and J. Moav. 1966. Profit in a broiler enterprise as a function of egg production of parent stocks and growth rate of their progeny. *Br. Poult. Sci.* **7**, 5-15.

Moav, R. and W. G. Hill. 1966. Specialized sire and dam lines. IV. Selection within lines. *Anim Prod.* **8**, 375-390.

Nemhauser, G. L., Rinnooy Kan, A. H. G., and Todd, M. J. 1989. *Handbooks in Operations Research and Management Science. Volume 1: Optimization.* North-Holland, Amsterdam, Netherlands.

Nicholas, F. W. and C. Smith. 1983. Increased rates of genetic change in dairy cattle by embryo transfer and splitting. *Anim Prod.* **36**, 341-353.

Niebel, E. and L. D. Van Vleck. 1982. Restricted selection indexes and selection response with overlapping generations. *Z. Tierzuchtg. Zuchtgsbiol.* **99**, 177-201.

Niebel, E. and L. D. Van Vleck. 1983. Optimal procedures for restricted selection indexes. *Z. Tierzuchtg. Zuchtgsbiol.* **100**, 9-26.

O'Neill, K. and L. D. Van Vleck. 1988. Potential of cytoplasmic effects of selection in dairy cattle. *J. Dairy Sci.* **71**, 3390-3398.

Owen, J. B. 1975. Selection of dairy bulls on half-sister records. *Anim. Prod.* **20**, 1-10.

Pasternak, H. and J. I. Weller. 1993. Optimum linear indices for nonlinear profit functions. *Anim. Prod.* **55**, 43-50.

Patterson, H. D. and R. Thompson. 1971. Recovery of interblock information when block sizes are unequal. *Biometrika* **58**, 545-554.

Pease, A. H. R., G. L. Cook, M. Greig, and A. Cuthbertson. 1967. Combined testing. pp 1-41. In: *Report DA 188.* Pig Industry Development Authority, Hitchin, Herts. England.

Peressini, A. L., Sullivan, F. E. and Uhl, J. J. Jr. 1988. *The Mathematics of Nonlinear Programming.* Springer-Verlag, New York, NY.

Pesek, J. and R. J. Baker. 1969. Desired improvement in relation to selection indices. *Canad. J. Plant Sci.* **49**, 803-804.

Petersen, P. H. and M. Hansen. 1977. Breeding aspects of embryo transplantation utilized in the bull dam path within a dual-purpose cattle population. *Livest. Prod. Sci.* **4**, 305-312.

Petersen, P. H., L. G. Christensen, B. Bech Anderson, and E. Oversen. 1974. Economic optimization of the breeding structure within a dual-purpose cattle population. *Acta. Agri. Scand.* **24**, 247-259.

Peters, J. A. (Ed.) 1959. *Classic Papers in Genetics.* Prentice-Hall, Englewood Cliffs, NJ.

Powell, R. L. and H. D. Norman. 1985. Trends in breeding values of dairy sires and cows for milk yield since 1960. *J. Dairy Sci.* **68** (Suppl. 1), 221.

Pym, R. A. E., and P. J. Nichols. 1979. Selection for food conversion in broilers: direct and correlated responses to selection for body weight gain, food consumption and food conversion ratio. *Br. Poult. Sci.* **20**, 73-86.

Ron, M., R. Bar-Anan, and G. R. Wiggans. 1984. Factors affecting conception rate of Israeli Holstein cattle. *J. Dairy Sci.* **67**, 854-860.

Ronningen, K. 1971. Selection index for quadratic and cubic models of the aggregate genotype. *Meld. Nor. Landbrukshoegsk.* **50**, 1-30.

Ruane, J. 1988. Review of the use of embryo transfer in the genetic improvement of dairy cattle. *Anim. Breed. Abs.* **56**, 438-446.

Samuelson, P. A. 1980. *Economics.* McGraw-Hill. New York, NY.

Schmidt, R. 1982. *Advances in Nonlinear Parameter Optimization.* Springer-Verlag, Berlin, Germany.

Searle, S. R. 1982. *Matrix Algebra Useful for Statistics.* John Wiley & Sons, New York, NY.

Shull, G. H. 1948. What is "heterosis"? *Genetics* **33**, 439-446.

Shultz, F. T. 1986. Formulation of breeding objectives for poultry meat production. *Proc 3rd World Congr. Genet. Appl. Livest. Prod.* **10**, 215-227.

Sivaragasingam, S., E. B. Burnside, J. W. Wilton, W. C. Pfeiffer, and D. G. Grieve. 1984. Ranking dairy sires by a linear programming dairy farm model. *J. Dairy Sci.* **67**, 3015-3024.

Smith, B. J. 1971. *The Dairy Cow Replacement Problem - An Application of Dynamic Programming*. Bulletin 745, Agricultural Experiment Stations, Institute of Food and Agricultural Sciences, University of Florida, Gainesville, FL.

Smith, C. 1969. Optimum selection procedures in animal breeding. *Anim. Prod.* 11, 433-442.

Smith, C. 1978. The effect of inflation and form of investment on the estimated value of genetic improvement in farm livestock. *Anim. Prod.* 26, 101-110.

Smith, C. 1983. Effects of changes in economic weights on the efficiency of index selection. *J. Anim. Sci.* 56, 1057-1064.

Smith, C., J. W. James, and E. W. Brascamp. 1986. On the derivation of economic weights in livestock improvement. *Anim. Prod.* 43, 545-550.

Smith, C. and S. P. Simpson. 1986. The use of genetic polymorphisms in livestock improvement. *Z. Tierzuchtg. Zuchgsbiol.* 103, 205-217.

Sokol R. R. and F. J. Rohlf. 1969. *Biometry*. W. H. Freeman. San Francisco, CA.

Soller, M. and J. S. Beckmann. 1982. Restriction fragment length polymorphisms and genetic improvement. *Proc. 2nd World Cong. Genet. Appl. Livest. Prod.* 6, 396-404.

Soller, M. and R. Bar-Anan. 1973. Breeding dairy cattle for meat and milk. pp 251-274. In: *Agricultural Genetics, Selected Topics*. R. Moav, Ed. John Wiley and Sons. New York, NY.

Spedding, C. R. W. 1975. The study of agricultural systems. pp 3-19. In: *Study of Agricultural Systems*. G. E. Dalton (Ed.) Applied Science Publishers Ltd. London, UK.

Strickberger, M. W. 1969. *Genetics*. The Macmillan Company, New York, NY.

Thompson, J. R., A. E. Freeman, and P. J. Berger. 1981. Age of dam and maternal effects for dystocia in Holsteins. *J. Dairy Sci.* 64, 1603-1609.

Thompson, J. R., A. E. Freeman, and P. J. Berger. 1982. Days-open adjusted, annualized, and fat-corrected yields as alternatives to mature-equivalent records. *J. Dairy Sci.* 65, 1562-1577.

Vandepitte, W, M., and L. N. Hazel. 1977. The effect of errors in the economic weights on the accuracy of selection indexes. *Ann. Genet. Select. Anim.* **9**, 87-103.

Van Tassel, C, P. and L. D. Van Vleck. 1987. Reasons for genetic gain being less than theoretically possible. pp 1-3. In: *Genetics Research 1986-1987 Report to Eastern Artificial Insemination Cooperative, Inc.* Ithaca, NY.

Van Vleck, L. D. 1981. Potential genetic impact of artificial insemination, sex selection, embryo transfer, cloning, and selfing in dairy cattle. pp 221-242. In: *New Technologies in Animal Breeding.* Academic Press, Inc.

Van Vleck, L. D. 1982. Is embryo transfer profitable? pp. 88-100. In: *Genetics Research 1981-1982 Report to Eastern Artificial Insemination Cooperative, Inc.* Ithaca, NY.

Van Vleck, L. D. 1986. Evaluation of dairy cattle breeding programs: specialized milk production. *Proc. 3rd World Congr. Genet. Appl. Livest. Prod.* **9**, 141-152.

Van Vleck, L. D. 1987. Reduction in genetic gain for milk production due to economic emphasis on test and type. pp. 23-31. In: *Genetics Research Report to Eastern AI Coop.* Cornell University, Ithaca, NY.

Weller, J. I. 1986. Comparison of multitrait and single-trait multiple parity evaluations by Monte Carlo simulations. *J. Dairy Sci.* **69**, 493-500.

Weller, J. I. 1988. Dairy cattle improvement in Israel. pp 50-73. In: *Proc. Poultry Breeders Roundtable.* St. Louis, MO.

Weller, J. I. 1989. Genetic analysis of fertility traits in Israeli dairy cattle. *J. Dairy Sci.* **72**, 2644-2650.

Weller, J. I. and E. Ezra. 1989. Optimization of a National dairy breeding program with multiple ovulation and embryo transplant for bull dams. *J. Dairy Sci.* **72**, Suppl. 1, 65.

Weller, J. I. and E. Ezra. 1991. Derivation of the March 1991 breeding index. *Meshek Habakar VeHahalav* **233**, 9-13. (In Hebrew).

Weller, J. I., and M. Ron. 1989. Trends in secondary traits in the dairy cow population in Israel. *40th Ann. Meeting EAAP.* **2**, 49-50.

Weller, J. I. and R. L. Fernando. 1991. Strategies for the improvement of animal production using marker assisted selection. pp 305-328. In: *Gene mapping: Strategies, Techniques and Applications.* L. B. Schook, H. A. Lewin, and D. G. McLaren (Eds.) Marcel Dekker, Inc. New York, NY.

Weller, J. I., H. D. Norman, and G. R. Wiggans. 1984. Computation of evaluations based on three parities as correlated traits by mixed model methodology. *J. Dairy Sci.* **67**, 2010-2020.

Weller, J. I., H. D. Norman, and G. R. Wiggans. 1984a. Weighing sire evaluations on different parities to estimate overall merit. *J. Dairy Sci.* **67**, 1030-1037.

Weller, J. I., I. Misztal and D. Gianola. 1988. Genetic analysis of dystocia and calf mortality in Israeli-Holsteins by threshold and linear models. *J. Dairy Sci.* **71**, 2491-2501.

Weller, J. I., M. Ron, and R. Bar-Anan. 1986. Multilactation genetic analysis of the Israeli dairy cattle population. *Proc. 3rd World Cong. Genet. Appl. Livest. Prod.* **9**, 202-207.

Weller, J. I., M. Soller, and T. Brody. 1988. Linkage analysis of quantitative traits in an interspecific cross of tomato (*L. esculentum* x *L. pimpinellifolium*) by means of genetic markers. *Genetics* **118**, 329-339.

Weller, J. I., R. L. Quaas, and J. S. Brinks. 1990. Multitrait analysis of a beef cattle herd with the reduced animal model. *Proc. 4th World Congr. Genet. Appl. Livest. Prod.* **13**, 390-393.

Weller, J. I., Y. Kashi, and M. Soller. 1990. Power of "daughter" and "granddaughter" designs for genetic mapping of quantitative traits in dairy cattle using genetic markers. *J. Dairy Sci.* **73**, 2525-2537.

Wiggans, G. R., I. Misztal, and L. D. Van Vleck. 1988. Implementation of an animal model for genetic evaluation of dairy cattle in the United States. *J. Dairy Sci.* **71**, Suppl. 2, 54-69.

Wilder, J. S. and L. D. Van Vleck. 1988. Relative economic values assigned to milk, fat test and type in pricing of bull semen. *J. Dairy Sci.* **71**, 492-497.

Wilton, J. W. 1979. The use of production systems analysis in developing mating plans and selection goals. *J. Anim. Sci.* **49**, 809-816.

Wilton, J. W., D. A. Evans, and L. D. Van Vleck. 1968. Selection indices for quadratic models of total merit. *Biometrics* **24**, 937-949.

Womack, J. E. 1987. Genetic engineering in agriculture: animal genetics and development. *Trends in Genetics* **3**, 65-68.

Wright, S. 1921. Systems of mating. I. The bioemtric relations between parent and offspring. *Genetics* **6**, 111-123.

Wright, S. 1952. Quantitative inheritance. pp 5-42. In: *The Genetics of Quantitative Variability.* E. C. R. Reeve and C. H. Waddington (Eds.) H. M. S. O., London, UK.

Yamada, Y., K. Yokouchi, and A. Nishida. 1974. Selection index when generic gains of individual traits are of primary concern. *Japan J. Genet.* **50**,

Index

Page numbers appearing in *italic* refer to figures and tables.

Lightning Source UK Ltd.
Milton Keynes UK
26 September 2009

144239UK00001B/29/A